THE CENTRE FOR FORTEAN ZOOLOGY
2004 YEARBOOK

Edited by
Jonathan Downes and Richard Freeman

Edited and compiled by Jonathan Downes and Richard Freeman
Cover and internal design by Mark North for CFZ Communications
Using Microsoft Word 2000, Microsoft , Publisher 2000, Adobe Photoshop CS.

First published in Great Britain by CFZ Press

CFZ Press
Myrtle Cottage
Woolfardisworthy
Bideford
North Devon
EX39 5QR

© CFZ MMVII

All rights reserved. Without limiting the rights under copyright reserved above, no part of this publication may be reproduced, stored in or introduced into a retrieval system, or transmitted, in any form of by any means (electronic, mechanical, photo-copying, recording or otherwise), without the prior written permission of both the copyright owners and the publishers of this book.

ISBN: 978-1-905723-14-0

CONTENTS

INTRODUCTION by Jonathan Downes	5
2003: A YEAR IN THE LIFE OF THE CENTRE FOR FORTEAN ZOOLOGY by Jonathan Downes	7
IRELAND'S EXOTIC FELIDS, by Neil Arnold	13
APPENDIX: The Country Antrim Cat-Flap Confusion! by Neil Arnold	31
INTERNATIONAL BIGFOOT SYMPOSIUM by Paul Vella	45
THE CURIOUS CAST OF THE SEAHAM SEA SERPENT by David Curtis	75
CRYPTO TV: Cartoon Cryptids, Abominable Ads and Monster Movies, by Neil Arnold	83
THE CUSWORTH CAT by Dave Baker	99
THE KNIGHT, THE CAT AND THE POET - The Tale of the Barnboro Wood Cat by Mark Martin	109
THE GREAT CROCODILE HUNT by Jonathan Downes	113
LOCH NESS: Steve Feltham Interview by Mark Martin	123
IN SEARCH OF GIANT BATS by Richard Freeman	127
CHUPACABRA: Jaimie Maussan Interview by Jonathan Downes	133
IN THE WAKE OF THE MONKEY MAN by Jonathan Downes	141
THE CRYPTOZOOLOGY OF DR WHO by Richard Freeman	169
MYSTERIES AND MONSTERS OF THE GREAT DEEP by Harold T. Wilkins - a reprint of rare booklet from 1948.	179

Jonathan Downes
(Director, Centre for Fortean Zoology)

INTRODUCTION

Dear Friends,

Welcome to the 2004 Yearbook.

Each year, whilst compiling the Yearbook, I am amazed at the sheer scope of subjects which we cover. What other organisation - for example - would put out a book containing an article on mystery bats, a long and slightly tortuous piece about urban panics in India, a scholarly examination of Mystery Cats in Ireland, and a piece of Surrealchemical looniness about how David Curtis invoked a sea-monster whilst getting drunk in his garden shed (which incidentally has been converted into a Jamaican theme bar)?

Of such diversity is - I believe - the CFZ made. It is because of this diversity that we have survived as long as we have, and will continue to survive well into the 21st Century.

Until next time,

Jon Downes,
Exeter
February 2004

2003: A year in the Life of the Centre for Fortean Zoology
by Jonathan Downes

Back: Graham Inglis, Jonathan Downes, Nigel Wright

Front: Mark Martin, Richard Freeman, Col. John Blashford-Snell, Mark North, John Fuller

-CFZ YEARBOOK 2004-

This, by anybody's standards, was a remarkable year. Indeed, it was probably our most successful year since the CFZ was formed back in the summer of 1992. I believe that, when we look back at the history of the Centre for Fortean Zoology in - say, another 10 years - 2003 will be seen as the pivotal year when, as Pete Townshend said: "This is when it all happened".

The year started spectacularly with our investigation into a series of zooform BHM Phenomena in and around Bolam Lake in Northumberland. It has always been a proud boast of ours that we are not fair-weather investigators. However, it takes something pretty spectacular to lure the four-man investigation team away from a nice warm office and up to snowy Northumberland in the middle of winter. Nevertheless, a spate of sightings of what most of the local media insisted on calling a Yeti was - by anybody's standards - pretty damn spectacular.

We achieved all of our mission objectives, including a mass sighting by me and no less than six volunteers, from what was then Twilight Worlds, the Research Group from Tyneside, who were seconded to the CFZ for the duration of the investigation.

The only bad aspect of this experience was the reaction that I, in particular, and the CFZ in general, got from large sections of the crypto-investigative community. In accordance with our charter, we published a preliminary report of the results of our investigations both online and in our journal. I was immediately vilified in many quarters by people who took exception to our explanation of events.

We stated that the phenomenon had to be Zooform in nature because a population of a hitherto unknown species of higher primate could not, under any circumstances, be living in a country park only 30 miles from Newcastle city centre. I still contest that this was a perfectly reasonable statement to have made. However, others did not agree, and we were attacked from all sides by people who wanted to believe that all sightings of such things are zoological in nature. I was banned from several Internet discussion groups as a result, and received a barrage of hate mail.

As a result, I was reminded - as if any reminder was necessary - of the classic axiom by the probably fictional Lazarus Long: "Never underestimate the power of human stupidity".

In April, we attended the Fortean Times Unconvention in London.

Richard and I gave a controversial presentation which drew criticism from some quarters. However we must have been doing something right because we gained over 60 new members and took over two grand. I would like to thank Mark North and Mike Playfair for working ridiculously hard all weekend manning the stall and giving out leaflets as Richard, Graham and I conducted a spate of media interviews.

-CFZ YEARBOOK 2004-

The CFZ in Sumatra
Dr. Chris Clarke, Jon Hare and Richard Freeman

The money that was raised at the Unconvention helped to finance our first major - and self-funded - foreign expedition. A three-man team, Richard Freeman and Jon Hare, and led by Dr Chris Clarke, spent over three weeks in the jungles of Sumatra on the track of the semi legendary ape-man Orang-Pendek. They returned with photographs of footprints and some hair samples. Sadly, their hair samples turned out to be from a cat species, but all was not lost.

As well as Orang-Pendek, these jungles are the home of another mystery beast - the lion-like cigau. Dr Lars Thomas, at the University of Copenhagen, has checked them against all but one known cat species from the region. The one remaining test to do is to compare them with the Asiatic golden cat. If he can prove that the hairs are not from this species, then we believe that we may have the first corroborative evidence of the cigau's existence.

Richard had only been back in the country for a few days when we were called away again. This time to a slightly less exotic location - suburban Staffordshire. There had been a number of sightings of a crocodile-like animal in a small pond. A team consisting of me, Graham Inglis, Richard Freeman, John Fuller, Nigel Wright, Mark Martin, Peter Channon, Wilf Wharton, Chris Mullins, and an observer from a northern News

Agency, spent several days hunting the pond and trying to track down an unscrupulous reptile dealer whom we believed responsible for dumping a spectacled caiman into this small and otherwise insignificant body of water.

Here, I would like to give my very special thanks and gratitude to Mark Martin, who not only carried out the initial investigations at no small cost to himself, but also made a number of a very generous and valuable donations of equipment to the CFZ.

In August, we were called in to help the Devon and Cornwall police investigate a particularly nasty animal mutilation case. The case has not yet gone to court and so we cannot give any more information at this time.

In October, we held our 4th annual convention - the Weird Weekend - at the Cowick Barton in Exeter. The star of the weekend was undoubtedly our Life President, Colonel John Blashford-Snell, but special thanks must go to Rachel Carthy who stepped into the breach at the last minute and delivered not one, but two excellent presentations. Other speakers were Richard Freeman, Nigel Wright, Tim Matthews, Adam Davies and Keith Townley, Mark North, Darren Naish, Chris Moiser and George Bishop. It was such a success - we raised over £600 in profit and about the same in donations of equipment - that next year's event will be held in the same location over the weekend of the 20th - 22nd of August.

In November, I visited the United States for the first time since 1999 and as well as appearing at a conference in Nevada, I made a point of liaising with Chester Moore - the CFZ rep for Texas. He amazed me with descriptions of his sterling work in various Cryptozoological fields, and I feel certain we shall be working closely with him in the future.
I would also like to thank Nick Redfern and his lovely wife Dana, for not only running the United States office of the CFZ, but for being such gracious and kind hosts during my stay. Also in November, I appeared at the 4th LAPIS conference in Blackpool, and Richard and I gave a well-received presentation to SPI (England) in London.

It has been a wonderful year for the CFZ. We have a hundred more members than we had a year ago, and I foresee no reason why that trend should not continue. I would like to thank John Fuller for his sterling work in the CFZ administration department and Elliot Saunders who, since becoming CFZ electronic information manager and corporate fundraiser, has made enormous strides towards making us the organisation that I have always wanted us to be. We are now - without a doubt - the largest and fastest-growing Cryptozoological research organisation in the world and I am exceedingly proud to be its director.

Jonathan Downes
Exeter
9th December 2003

IRELAND'S EXOTIC FELIDS
by Neil Arnold

- CFZ YEARBOOK 2004 -

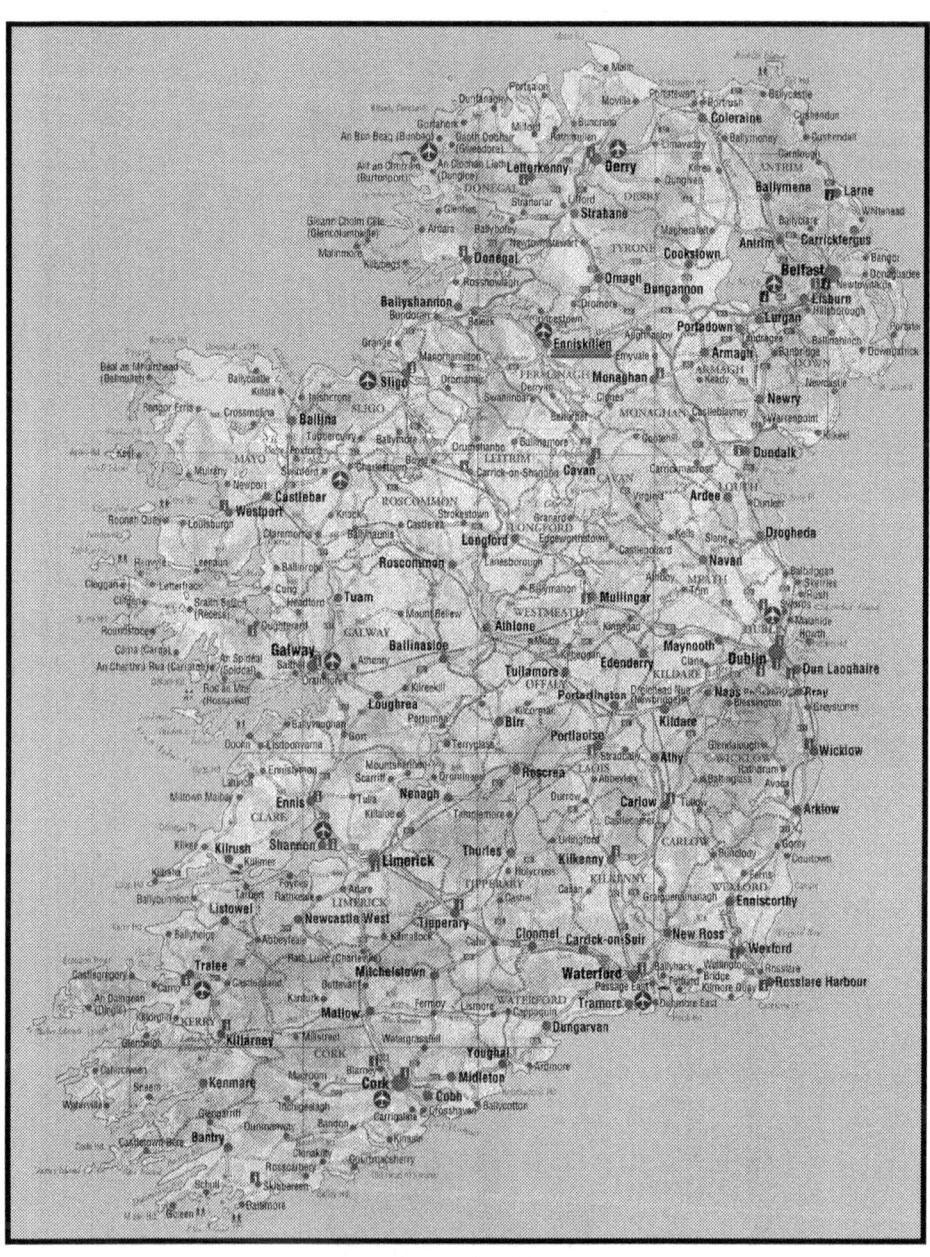

Map of Ireland

Neil Arnold runs KENT BIG CAT RESEARCH, studies folklore and publishes his own material, including two books on zooform phenomena, Odd-Bodies & Clandestine Creatures. His own exotic cat research can be found at: www.tudor34.freeserve.co.uk/beast_of_blue_bell_hill.htm

On 15th February 1996 it was claimed that several witnesses had sighted a 'lion cub' in Fintona, Co. Tyrone, Northern Ireland. Three days later the felid was dead, shot by marksmen from the Royal Ulster Constabulary (R.U.C.), after it was seen slinking close to a field of sheep. The so-called 'lion cub' was nothing of the sort, but still very much an exotic cat that had obviously escaped, or been released from a private collection due to the fact that it was wearing a collar around its neck.

Several reports on the Internet identified the specimen as a caracal, although the photo clearly shows that the felid is a Eurasian lynx, due to the shortness of the tail, and the ears certainly do not match the size of the caracal. The incident itself increased demands across Ireland for the introduction of the Dangerous Wild Animals Act, which, across the rest of the UK, was introduced in 1976 to stop such escapes occurring and to require those who kept such felids in captivity to have an official license. As for the carcass of the lynx, contact was made with the R.U.C. who stated that the corpse of the felid would be stuffed and displayed for the R.U.C. museum. Further investigations mounted by the Centre for Fortean Zoology, though, indicated that the body would eventually be destroyed.

Large cats such as black leopard, puma, and lynx have been enclosed within inadequate facilities for a number of years, and these felids remain a separate enigma to the reports of the Irish 'wildcat' that is said to exist. It is highly likely that a majority of the felids, of a larger nature sighted across Ireland, are cats that have escaped from menageries, or were let go on purpose. The mystery itself does not necessarily pose the question as to how they got there, but rather for how long they have been around.

In 1928, in Co. Derry, a Mr. Martin was trout-fishing near his rural home when he saw a large, black animal strolling through the shallow water of the river. Unnerved by the nearing presence, Mr. Martin, realising it wasn't a dog, leapt up onto the bank, startled by the menacing creature. Mr. Martin hid behind the nearest tree, and as the animal padded by it gazed eerily in his direction, it bared its teeth and stared menacingly with glowing eyes which, at the time, were described as 'blazing red'.

The animal went on by, leaving the witness terrified, as he remained cowered amongst the leaves of the tree. When the animal had disappeared from view, Mr. Martin hastily made for home, unsure of what he'd seen in the water. Days after the encounter, Mr. Martin made a few enquiries in the local area to see if anyone else had sighted the animal, but his research drew a blank, until one day he stumbled upon the legend of the Irish Pooka on a cigarette card. The animal on the card appeared as a giant 'fairy dog', with a black coat and fiery eyes. Mr. Martin decided to hunt the creature, but he failed in his attempts to track the beast down, a beast often reported during the hours of dark-

A lynx shot in Fintona, Co Tyrone, Northern Ireland.

ness by walkers and fishermen on the water, but hardly ever by daylight.

Such encounters with large, mysterious black animals are common in the folklore of many countries across the world. They are known as Black Dogs, and many encounters with these ghostly hounds obviously suggest supernatural connections, but there is also the possibility that encounters such as the one involving Mr. Martin do involve mystery cats such as the black leopard, despite the fact that the legendary Pooka is often tied to watery locations - but then again, so are cats!

The incident from the 18th February, regarding the shot lynx, may appear far removed from the 1928 encounter involving Mr. Martin in Co. Derry, but what these cases prove is that there has always been an abundance of strangeness in Irish lore regarding felids - and indeed, such lore is common throughout the world - but let us not put these cats into the realm of the supernatural which is haunted by the strange hellhounds.

An Arctic wolf was shot dead after eight days on the loose during the Autumn of 1995, in Lisnaskea, Co. Fermanagh. On the 29th of June 1997 Mark Rowe, writing for the Independent On Sunday, in an article entitled Power Pets – The New Love Of Ireland, expressed the concerns of exotic animals being kept in captivity, with no licensing:

"There may be no snakes in Ireland but there are big cats galore. Tigers, pumas and jaguars are fast becoming status pets in the north and the south. Ray Cimino, of the Dublin-based Trust For The Welfare Of Captive Wildlife, said: 'It is a growing problem. During the past five years the number of people owning tigers and other big cats has multiplied several times. It is not unusual to see them being taken for walks down country lanes. The R.U.C. has had complaints that tiger cubs on leads have been taken into shopping centres.'

The reason for the ownership of exotic pets - bears too have been found by officials - is, as already remarked, that the 1976 Dangerous Wild Animals Act does not apply in Northern Ireland and no similar legislation exists in the Republic either.

Angi Carroll, deputy director of the Dublin Society for the Prevention of Cruelty to Animals said: 'In Ireland you must have a license for a dog but there's nothing to stop you walking down the High Street with a tiger or a rhino. Most pets are sold in Belfast so people can drive up from Dublin and back across the border with a tiger cub within four hours. There is nothing furtive about it.'

Ms Carroll and her colleagues can act on reports of wild animals being kept as domestic pets only if they believe the creatures are housed in poor conditions. Then they can prosecute under the 1911 Protection Of Animals Act.

Last year, police on a routine drugs raid in a Dublin suburb found an adult jaguar and a serval in a garage. In another incident at a Limerick farm, an ex-circus bear, two Siberian tigers, two tiger cubs and a baboon were found. The owner was prosecuted and banned for life from owning animals.

Many of the animals are sold by unscrupulous zoos or by travelling circuses. Others are bred in captivity. A tiger cub can be bought for as little as £150 and sold for £2000.

However, many of the cubs do not survive into adulthood. 'People think if they hand-rear a cub it will grow into a well-trained adult, but these cubs are taken off their mothers at a very early stage and often die by the age of five months because they haven't had the right nutrition', Mr Cimino said. 'People have no knowledge of how to raise them. They are not prepared to spend £25,000 on a proper secure habitat, and so put them in garages or small cages in back gardens.'

Ms Carroll said, 'At the moment these creatures are freely advertised in Dublin newspapers. We urgently need a change in the law but I suspect nothing will be done until somebody gets mauled.'

The Royal Ulster Constabulary's wildlife liaison officer, Inspector Mark Mason, believes that an Act should be introduced to control the ownership of wild cats, but he also wants any such pet to have a microchip inserted in its neck to make it easily traceable."

On September 28th 1999, the Belfast Newsletter reported that the Ulster Society for the Prevention of Cruelty to Animals (USPCA) had warned the public to be on the lookout after several sightings of the 'beast' of Ballygowan, Co. Down. A felid, measuring between four and six-feet in length, black in colouration, had been sighted near Davidson's Quarry and also at a school playground on the Belfast Road. Local welfare officers believed that the animal in question was an escaped 'panther' (although the term 'panther' is, across the USA, used to describe the fawn-beige coloured puma, in Britain it is used to describe the black leopard, which is often called a black panther).

An officer from the USPCA, Martin Dobbin, claimed to have found unusually large prints in the ground near the quarry, and also said that if he saw the animal he would call in the vets from Belfast Zoo to tranquillise it. Nothing came of his investigation.

During '99 there were several reports of large exotics filed. *Fortean Times* magazine put together a short list of reports for the year, with sightings from Aughnacloy, Co. Tyrone - puma (1st January), Sixmilecross Co, Tyrone – brown cat (2nd January), Black Mountain, Belfast - large cat with small dog in its mouth, (late Jan), and a further twelve sightings of a large black cat near Ballygowan, and even a report of an escaped red panda, said to be inhabiting Black Mountain.

During the Spring of 2000, David Malone, executive producer of the film company Extreme Production, contacted various newspapers etc, in the hope of bringing forward witnesses to big cat activity across Ireland, and also to hear from anyone who kept an exotic cat. His research and investigative results were to be aired in a Channel Four documentary regarding Northern Ireland and its situation.

Reports over the millennium were reasonably quiet for an island where large animals are kept in shoddy conditions! During 2001, large cats were seen at Richhill, South Armagh and further south, in February at Rearcross, Tipperary, and in West Wicklow during July, there were two very peculiar reports of large cats. A Rearcross witness claimed that the animal they saw was striped with a bushy tail, whilst at Blessington, West Wicklow, the cat sighted was described as being 'cheetah-like'. My own exotic cat research in Kent, England, has previously enabled me to siphon out the hoaxes from the genuine, and the possibilities from the impossibilities, but across Ireland it is proven that anything is obtainable, which means anything can escape into the wilds. One cannot rule out some reports of felids simply because witness descriptions are all too varied, and sometimes exaggerated. However, if alleged sightings of 'cheetahs' do not resurface time and time again then one has to question their validity, although it is fair to say that smaller cats such as the jungle cat, and in some cases the golden cat, can disappear from public view and so reports will be scarce, making it nigh on impossible to build a better picture of the felid in question, especially concerning its territory.

On Tuesday, May 22nd 2001, the Irish Examiner, reported: Farmers on alert after big cat goes on the prowl.

Joe Oliver reported: "The RUC has warned farmers on the Border to be on their guard after it was confirmed that a puma or a panther is on the loose. There have been a number of sightings of the beast in South Armagh in recent weeks. Evidence of animal droppings at a remote house also pointed to the presence of a large cat.

R.U.C. Wildlife liaison officer Mark Mason said teeth marks were found on a rabbit in a hutch in Co. Down. Mr. Mason ruled out any possibility that the bite marks came from a dog. "The distance between the canines is consistent with a cat of the scale of a puma or a panther," he said.

"It ripped the rabbit hutch apart and all that was left was a spot of blood. The fact that in recent times we shot and have recovered a lynx proves not all of these sightings of big cats are untrue. There is firm evidence to support the view these sightings are not all fairy stories."

In 1996, police marksmen shot dead an African lynx (author's note: incorrect continent stated, the lynx was European - the caracal is native to Africa but the photograph clearly shows a lynx) in the Fintona area of Co. Tyrone.

Two years ago dozens of locals, including doctors and soldiers on duty, reported a puma-like animal roaming the countryside.

Dubbed 'The Border Beast', it was sighted in Augher, Clogher, Ballygawley and Aughnacloy.

Mr. Mason said the beast which destroyed the rabbit hutch would be unlikely to attack humans.

"I would advise people to be on their guard while it remains at large", he said.

Various liaison officers and so-called 'experts' within the wildlife field often speak of these so-called 'panthers' and 'puma-like' felids, but seem to have trouble identifying these animals despite countless reports of them. Let us then look at the candidates:

BLACK LEOPARD (*Panthera pardus*)

A common novelty cat kept by owners during the swingin' Sixties. This is merely a darker-coated version of the spotted leopard, growing to over four feet in length, standing over two feet at the shoulder and weighing over one-hundred pounds.

The leopard is a highly agile felid that can climb trees, especially in order to stash prey so that other predators in their countries of origin (Africa/Asia) are unable to feast on the kill they have made.

Such felids live for around fifteen years in the wild, and are solitary hunters, but up to four young can be produced. Black (melanistic) leopards can only produce black offspring but normal leopards can have a mixed litter.

The leopards sighted across Britain are always black and it is highly unlikely that black pumas are being seen, as they are extremely rare despite the fact that across the USA a number of witnesses describe black cats thus. They probably tend to use the term 'black puma' because the puma (cougar) is the native felid.

The black leopard is a powerful cat, sleek, muscular and slender. There is enough prey in the British countryside to support a population, and whilst livestock often suffers, rabbits are the most ideal prey, alongside deer, rodents, birds, and foxes.

Across the UK, black leopards appear reasonably common and are the main felid sighted, although the puma is also well established.

PUMA (*Felis concolor*)

The puma (aka: cougar, mountain lion, panther), is native to the US, although endangered in parts, and despite growing to a size that compares to the leopard, if not bigger, it is not a 'big cat' per se, as it purrs - in the same way as members of the genus Felis - and cannot roar. The puma also screams rather than roars, this particular is the largest of the lesser cats.

Many people across Ireland have described encounters with a 'black, puma-like' animal but the puma is not black in colour: its coat can vary from silvery-grey to fawn-beige and there are no markings although young are born spotted.

Attacks on humans have occurred across the USA, but other creatures such as deer, dogs and bees cause more deaths each year.

EURASIAN LYNX (*Lynx lynx*)

The Eurasian lynx is twice the size of its North American relative. This felid is known for its leggy appearance, shortish tail, triangular ears topped with tufts of black hair, and well-furred, large paws. The coat can appear in a variety of colours from yellowish-brown, to reddish-grey with dark spots. These cats are often found in rocky regions and forest areas, their natural environment covering western Europe, the former USSR and parts of Asia. The Spanish lynx (lynx pardinus) is similar in appearance and is sometimes classified as a subspecies rather than a species in its own right.

Many people confuse the caracal (lynx caracal), with the lynx. The caracal, native to Africa, and India, is smaller than the lynx, is reddish brown with long, pointed ears with black tufts. They are excellent jumpers, known to pluck low-flying birds from the air, and they are solitary hunters. Felids such as the caracal, jungle cat (felis chaus), and more so the Asian golden cat (felis temmincki), could well be incorrectly identified by witnesses, across Ireland and the rest of Britain. The Asian golden cat, could, at a brief glimpse to an unsuspecting witness, appear puma-like, whilst a puma on the horizon could appear as a black leopard.

With so many private menageries spilling into the woodlands, all manner of felids could be roaming the Irish woodlands, fields and towns, especially at night when most of these cats prefer to feed. Smaller cats such as the caracal and jungle cat can disappear without trace, feeding off mice, lizards, insects, birds, and remain undetected for years, but very rarely are these felids identified in the press, who merely seek a 'beast' story to make their headline.

On Tuesday, October 29th, 2002, LAOIS NATIONALIST reported, Another sighting of panther, with an article that not once accurately described the felid!

"There has been another claim of a sighting of the mysterious cat-like animal in the county.

Prompted by the article in last week's Laois Nationalist about claims of a sighting of a panther in Mountmellick, a woman caller rang to tell us about her encounter.

For fear of ridicule, the caller asked her name to be withheld: '…it was on Monday morning about 9:10 when I was on my way to work in Kildare. I was approaching the cross at Ballybrittas on the Main Dublin road when what I thought was a big black dog bounded out in front of the car and bounced across the road and headed down the Vicarstown Road.

I wondered what type of dog it was. It had a jet-black coat and I could plainly see it had a tan-coloured collar, about three-inches in diameter, around its neck.

It was on my mind all day and I wondered who would have come off worse if I had been a few seconds earlier. It didn't look like a dog running about after being out all night.'

The caller said she was on her own in the car but there was other motorists on the road at the time and they must have seen the animal.

'It wasn't until I read the paper on Wednesday, that made me think again, that maybe it wasn't a dog. It seemed to have been well built. I'm not sure how it was a dog, it could have been the panther that was spotted in Mountmellick', she said.

The area the woman caller claims to have seen the animal is approximately eight miles from where two men, John Travers from Mountmellick and his brother, claim they saw a panther the previous night.

While the Mountmellick Garda say they received no reports of dead animals, and are unaware of anyone keeping a panther, or similar big cat, they have not ruled out the possibility of someone owning one locally."

Melanism is common amongst a variety of species of felid, but the black leopard makes up 95% of eye-witness sightings which describe black cats, although in Scotland the Kellas Cat has come under investigation. Cats such as the leopard and puma have vast territories, up to 250-square miles in their countries of origin, but across Britain a male may have a territory up to 100 square miles and this area may cover the territories of several females. The Mountmellick cat is quite obviously a black leopard, but across the UK the large cats sighted appear to be British leopards and not cats which have escaped from captivity over the last ten years or so. Until laws are introduced into the Irish system to curb them, private collections in poor condition will continue to spill into the forests, but even if an act is passed to prevent such easy access to exotic cats, there may already be a well-established population of puma, leopard, lynx and others, which in turn will become more abundant over the years, in the same way it has snowballed in Britain.

On 1st March 2002, Peter Gleeson reported for the Nenagh Guardian, about the **'SEARCH ON FOR CREATURE WITH FANGS AND CLAWS'**, after several reports near Rearcross of an unusually large felid too big to be a domestic cat. He wrote:

"The Scottish wildcat, is there one in Rearcross? Yes, the search is on for a genuine wildcat suspected to be wandering around the remote mountain forests near Rearcross.

"It's wild, with claws and fangs. And it is believed to be roaming the countryside around North Tipperary.

"This is not your ordinary tabby. No, it's bigger, with a striped bushy tail, and it's not supposed to be native to this country at all.

"The search for the wildcat started after Sandra Garvey, who lives in the area, spotted an animal she had never seen before when out driving one night just over two weeks ago.

"Ms Garvey, who lives in the town land of Knockfune, told The Guardian, 'I nearly drove off the road I was so shocked. I saw this thing and what was striking about it was its tail. I saw nothing like it before….it was larger than your average moggie.'

"When Ms Garvey went home and related the strange story to her husband he confirmed that he had seen a similar animal a year previously.

"Ms Garvey's daughter told the story to her teacher, Jeff Griffin, Vice Principal at Villiers Secondary School in Limerick, and a regular walker in the area, who confirmed that three years ago he too had seen an animal that fitted the description.

"Ms Garvey contacted the RTE radio nature programme, 'Mooney Goes Wild', to tell her amazing story to the nation last weekend. Even the shows presenter Derek Mooney, travelled to Knockfune to search for the elusive wildcat - alas without any success.

"When contacted by this newspaper earlier this week, Derek Mooney said there was no firm evidence of the presence of wildcats in Ireland. The species was found in Scotland, Europe and Northern Asia. 'There is the possibility that someone brought a wildcat into the country,' said Mr Mooney. Another possibility was that offspring of once domestic cats had turned feral and appeared wild looking.

"'The chances of it being a genuine wildcat are very slight from what I have been told by scientists. You have to prove it and what we are looking for is that if someone sees this creature again they should take a picture.'"

The Guardian contacted an expert on the native Scottish wildcat, Alan Paul, who confirmed that the species had never been native to Ireland. The only way the species could have got here was if they had escaped from a zoo or some other form of captivity.

'It's very rare for anyone to see a wildcat. They like to keep out of harms way.'

Mr Paul said the animal was not a danger to humans, unless provoked.

'It will attack you if you corner it. It's no more vicious than any other animal of the same size.'

Jimmy Greene, a wildlife ranger in North Tipperary up to two years ago, but now patrolling in Laois and Offaly, believes wildcats exist in Ireland.

Jimmy said he had seen a wildcat and its kittens while patrolling in the early hours in Slieve Bloom mountain range in Offaly. 'I knew straight away it was not an ordinary cat. It was pure wildcat. You have to be up early in the morning or at night to see them. I didn't think we had them in Ireland before that, but there are also reports of them in County Wicklow.'

Jimmy said he never saw a wildcat during his years as a ranger in North Tipperary. He believed the animal that Ms Garvey had most likely seen was a pine marten, which, he said, also had a bushy tail and were not much bigger than a wildcat.

Ms Garvey is still convinced that what she saw could very well have been a wildcat - but maybe not of the genuine kind found in Scotland and parts of Europe and Asia.

'Over the years we have had feral cats in this area, but they have grown larger and developed into an indigenous wildcat. They have been interbreeding over the years.'

In his book The Smaller Mystery Carnivores Of The West Country (1996), fortean zoologist Jonathan Downes writes, "Although it is a common figure in Celtic-Hibernian folklore, the wildcat has never, officially been recorded from the island of Ireland. There have however been isolated records of what appears to be genuine Irish wildcats for centuries."

Sub-fossil remains were found in two County Clare caves during 1904, and investigating doctor, R.F. Scharff found that remains were more comparable with the African wildcat instead of the European species! Whilst some try to explain away wildcat reports as pine martens, according to Karl Shuker in his superb 1989 book, Mystery Cats Of The World, inhabitants of the hills of Kerry knew both the pine marten and 'wildcat' as separate species, with the pine marten often being called the 'tree cat' and the wildcat 'hunting cat', so there is no reason at all to dismiss any modern reports of the felids that are not supposed to exist in Ireland. Of course, there is always the possibility that some of these wildcats have been imported and released/escaped into the wilds, but many reports of these cats do date back several hundreds of years.

According to Shuker, wildcats were once caught in a trap set up by a gamekeeper during the 1800s and there were also captures during the 1900s; large cats weighing over 10 lb (5 kg) had been sighted around Mayo, and one specimen had been shot and killed in County Antrim. Many have offered opinions that the wildcats loose throughout Ireland are hybrid forms, i.e. wildcat crossed with domestic cats, but as Shuker correctly states, "....there must be pure-bred wildcats there in the first place." Others have surmised that remains found in caves of wildcats relating to African species are nothing more than remnants of Egyptian mummified cats imported across Western Europe during Victorian times. Felids were often ground down to produce fertiliser. However, whatever the truth behind the legend and lore of the Irish wildcat, it is something that will always continue, despite many suggestions that any kind of felid is probably extinct. But so long as witnesses such as Sandra Garvey continue to see mystery cats that appear to resemble the mythical wildcat, the saga shall continue.

I have never been one to connect the lore of ghostly black dogs to flesh and blood sightings and evidence of exotic felids, however, throughout the world there are reports of Black Dogs that seem to relate all too closely to black leopard reports instead of anything remotely paranormal, whilst others, although anecdotal, could also suggest mystery felids rather than hellhounds, especially when we consider reports of historical value from over one-hundred years ago. This of course is not a way of dampening the fiery-eyed black dog enigma, and certainly will not as many witnesses to these unearthly dogs have been close enough to see an animal that is most definitely not like a cat, and does display supernatural qualities, mainly to the extent that they disappear.

During the early 1900s a massive 'black dog' was sighted near Ballaghadereen in County Roscommon, Ireland, by a witness who described the animal as being, "....as tall as a person's shoulder", and which seemed to "....disappear through a closed gate." In 1913, as described in Janet and Colin Bord's superb Alien Animals, a school-master residing in Ballygar, County Galway, spoke of an eerie encounter with a black animal which followed him along a quiet road at dusk one evening. The school-master was cycling to town when he saw the animal, which he described as a 'black dog' which made him feel uneasy. The animal allegedly then stopped and let the worried witness continue his journey.

During the year of 1874, something clearly not supernatural, or dog-like either, was stalking sheep at Cavan. According to sources, and the files of Charles Fort, who wrote of the incident in his book LO!, some thirty sheep were killed during a one night raid. Rather oddly, no flesh was devoured, but the sheep bore throat punctures and were bereft of blood. Whatever slaughtered the sheep left long, elongated tracks, and a similar individual also seemed to be on the prowl during the same year, but one hundred miles away at Limerick. The Weekly News of Cavan reported that several people were attacked by the mystery marauder, and it was alleged, in a story dated April 17th of 1874 that some witnesses to the 'beast' were admitted to an insane asylum after encounters with the creature! Of course, and as Charles Fort correctly pointed out, "…Damn the dearth of details in the Irish and British Press! Journalists of that period had a frustrating way of writing ambling essays which only hinted the facts."

Interestingly, reports of other similar kills across Britain were quite common at the time, certainly to suggest that some populations of cat-like animals were indeed prowling the British wilds, even if not all the press reports could accurately describe cat kills. With so much lore across the world pertaining to the enigmatic powers of animals, whether as mystical creations, portents of death or simply elusive, majestic wonders, many historical accounts remain almost undecided, tales that could be of great use to exotic cat researchers if only the facts could be siphoned from the fiction.

In the nineteenth century, as recorded in Lady Gregory's Visions & Beliefs, (Volume 2), something black, the size of a calf, was seen in an unspecified location in Ireland. The creature, originally believed to be dog-like, despite the fact that there is no record to suggest so, also bore some kind of illumination around or in its mouth, and as it loped across a field the sound of chains were heard, a feature often described in Black Dog lore although I have often considered the possibility that if such an animal was an escaped black leopard, then the chains could well be hanging from its neck after it had broken free from its manacles! Another report from Lady Gregory described an eerie road encounter with a large beast. The incident took place in Kinvara, Co. Cork one night when an animal appeared on the left side of the horse-drawn car, and then vanished from sight as quickly as it had appeared.

A big, black animal was also sighted in the 1950s near Derrygonnelly, Co. Fermanagh, this time by what would be considered by sceptics to be a credible witness in the form of a police sergeant. He was on his cycle one evening, riding near a crossroads through the tranquil village, when he caught sight of a large, black animal, thought to be a dog, which kept pace with him on his journey. The sergeant claimed that the animal seemed to glide effortlessly, and had no trouble keeping up with him despite his sudden burst of speed. The most eye-catching characteristic of the animal was its large and 'glittering' eyes that filled the man with dread. As the sergeant reached his destination, the animal was nowhere to be seen. He immediately told his comrades of the encounter who all came to the conclusion that the creature was an omen of ill-luck, and by coincidence, the police sergeant fractured his leg shortly after the encounter.

Entities such as the Black Dog and the Banshee have often been considered portents of doom, and symbols of misfortune through lore, although the black, graceful gait of the so-called 'black dog' and mournful cry of an unseen banshee could well have their origin rooted firmly in the mystery we have come to know across Britain as the 'alien big cat' saga. A haunted house in Co. Antrim was often said to be plagued by the mournful cries of a banshee spirit that lurked in the woodlands and rolling fields nearby. Despite the ghostly activity within the house, which is of no relevance here, one happening of strangeness does seem eerily characteristic of a prowling 'big' cat. The peculiar noises heard by several witnesses, had come to be known as the Invermore banshee. A resident, a Miss Thompson, witness to various paranormal occurrences, heard the awful wailing one night as it echoed around the garden in the darkness. The sound ended as a terrible, haunting shriek and she commented that whatever made it must have had a massive pair of lungs, as it howled through the night. Although a death in the family occurred shortly after, one can always connect the strange with the strange!

The weird shrieking though continued for a long time, always soaring over the valleys and resounding around the grounds of the house. Miss Thompson described a second incident that awoke her one night as, "…ranging from a roaring to an eldritch shriek." Two small dogs owned by the family strangely remained asleep during the cries, although another witness, a Miss McClane, said, "Our hair quite literally stood on end." The women, when questioned by author Sheila St. Clair, said they were unsure if the cries were human or animal, but it was certainly like nothing they had heard before. The women also claimed that the noise was so eerie and piercing that it seemed as though it was in the room with them.

As mentioned earlier, the puma (mountain lion, cougar), does omit a very eerie cry-cum-shriek that can travel for miles and would certainly be a sound that witnesses would not be familiar with. It would be fair to say that if anyone not familiar with the cry of an out of place large cat heard such a noise, they would be terrified and certainly look towards the eerie banshee wail for a solution, in the same way witnesses to so-called black dogs may consider paranormal connotations.

A large, black dog-like entity is said to haunt an old bridge over the River Quoile, at Downpatrick, Co. Down. Many theorise that the creature sticks to stretches of roads and waterways as some kind of guardian, and a similar black animal has also been sighted near a bridge at Pontoon, Co. Mayo, but unless witnesses describe accurately something more akin to a phantom hound, I am still more inclined to believe that what these legends suggest are escaped/released black leopards, which often use railway lines, bridges, waterways and roads to navigate their routes. It is the Black Dog seen inside houses etc, which seems to point more towards paranormal forces. Such 'dogs' are divided into three types:

1) The Barguest, a shape-shifting demonic entity,

2) The Black Dog, a calf-sized, uniform in type dog, pure black, sometimes shaggy, and

3) a more scarce type which seems to appear in conjunction with particular days of the calendar.

However, during the 1700s, and right up until the 1950s, if anyone came face to face with a long, muscular creature in the night crossing the road, then surely, judging by so many other experiences, it is most likely that what they saw was a prowling melanistic leopard rather than a phantom hound, despite the lore of the black dog. It's certainly confusing. Stranger still is the fact that in the modern era Black Dog reports have diminished rapidly, and sightings of large, exotic cats, particularly big, black ones have come to the fore. Are we merely seeing a transformation of lore within the belief system or a natural mystery simply taking on its true shape, a shape which has been ignored for centuries as terrified witnesses described seeing phantom, black dogs with glowing eyes when the reality was, they were seeing gracile cats with reflective eyes slinking into hedgerows like ghosts.

On a final note, I would like to end the intriguing mystery of Ireland's mystery felids with an interesting account that first appeared in Issue 27 (2002) of Animals & Men magazine, which describes, "…an encounter with a Pooka". (The Pooka is an Irish entity said to take on many forms, often though as a dark coloured, horse-sized phantom) The story was submitted by reader Louise Donnan.

"One clear summer evening in 1997, myself and my niece Claire (aged 17) went for a run in the car as Claire had just passed her driving test. We went out to the Grayfield-Greencastle district, four miles outside of the town of Kilkeen, Co. Down. We were travelling along a long, straight stretch of country road when we both spotted, in the distance, ahead of us an animal on the grass verge besides the road. From this distance it looked like a sheep, but we both agreed that it seemed like a very large 'sheep' indeed with a coat made up of what looked like bits of torn rags as opposed to wool.

As we approached the animal we slowed down to get a look at it, as we were both bewildered to what type of animal it was. Just as we were adjacent to it, it turned its head and looked directly at us. We both gasped in disbelief and revulsion. I no sooner had the words, 'What the hell is that?' out, when it charged straight into the car which at this stage was almost at a stop. Its face was right up to Claire's window and both of us - just for a second or so - looked right into its eye. I say 'eye' as the other eye was covered by its tatty coat.

We were both almost frozen in fear as the eye looking straight at us was reddish in colour, and gave a terrible wild penetrating stare. When I looked in its eye I could almost see its mind working powerfully behind it, a mind not of an ordinary animal but one with another sense of evil which I had never encountered before or since. I felt sick with fear, but thankfully Claire was able to compose herself enough to accelerate the car and we took off at an impressive speed. Our relief was very short lived as we suddenly felt a 'thud' at the side of the car. To our horror this mad 'animal' we thought we had gotten away from, was running alongside us and deliberately banging into the car.

I screamed at Claire to go faster which she did, and we both felt terror and disbelief that this 'thing' was able to keep up with us. Just as I felt that we weren't going to get away, the 'animal' suddenly stopped the chase and just stood in the middle of the road watching us as we escaped at great speed. As I looked behind me out of the rear window of the car I got the impression that the 'animal' hadn't even tired, but had, for some reason, reached the decision to go no further as it had reached the edge of its territory. It was some time and distance later before both Claire and I felt safe from it and we were both still saying things like, 'What was that ?', and, 'How could that thing travel at that speed ?'.

When we returned home we were both very shaken by the experience and told family members what had happened. Naturally we were told that we had overactive imaginations and that it must have been just a large dog, but we both knew what we had seen was no dog but abandoned the subject as we knew we weren't going to be believed and perhaps a bit of fun-making may have started and neither of us were in the mood for this….

The last time I spoke to Claire about it we both sat shaking our heads and the conversation ended…

Claire: That was no dog.

Louise: That was no sheep.

Claire: That was evil…"

Fellow researcher and good friend, Gary Cunningham who lives in Ireland is related to the witnesses. He told me in a long phone call that although the details were clouded and possibly exaggerated, he is sure that what the women saw was something very unusual.

Judging by such confused yet clearly terrifying accounts, there is no telling as to what strange animals, flesh and blood or otherwise, are roaming Ireland's lonely back roads and dense woodlands. Whilst some are firmly embedded in the darkness or lore, others have merely escaped from menageries, and are establishing themselves in the wilds to become a native species. Such animals are at present ignored by authorities, and so may tragically drift into a mythical status. However, populations of these felids are clearly rising in number, and there has to come a time when they will slink out of the depths of the unknown and into one too many back gardens, making their presence felt as Ireland's top predator, days that locals thought were over when the last wolf was wiped out.

It is certain that several unidentified species of animal lurk within the murky waters of Ireland's lakes, and foggy lanes, but the felids that remain so elusive are obviously there, but they will always remain mysterious and take on different forms when those who see them have been deprived of the truth for so long by authorities etc; the truth that Ireland is very much 'wild' cat country, in every sense of the word.

However, if you still remain unconvinced that large cats are roaming the woodlands of Ireland, or ever wondered how some of these cats got there then I'll leave you with this story, from the middle of July 2003 that was posted on the Centre for Fortean Zoology website forum:

A JAGUAR IN THE GARAGE

by Toby Harnden, Ireland Correspondent.

"Keeping a jaguar in the garage would not normally upset the neighbours – except, perhaps, when it is a large carnivore from the Amazonian rainforest. Garda officers searching a semi-detached house in Clondalkin, Co. Dublin, discovered a fully grown female jaguar in the garage.

Having expected nothing more fierce than a Ford Escort, they beat a hasty retreat and radioed for help.

An inspector from the Dublin Society for the Prevention of Cruelty to Animals arrived with a tranquilliser gun. As well as the jaguar, called Princess, he found a serval.

A Garda spokesman said: 'The lads got quite a fright when they opened the garage up. It will be a question of dealing with the matter under the Wildlife Act rather than a motoring offence.'

The owner of the jaguar said he took it out for walks at night and fed it on pigs' heads. Both cats were taken to Dublin Zoo.

Maurice Byrne, the DSPCA inspector, said the jaguar was kept in a wooden crate. The Garda said the owner, who claimed he was, '...looking after the serval for a friend', may be charged with not having a license for the jaguar.
The creature's future is looking uncertain. The head curator at Dublin Zoo said: 'We already have two jaguars.'

With thanks to:

Gary Cunningham, Jonathan Downes, Karl Shuker, Scottish Big Cats, various Irish papers, and especially Nick Sucik for giving me the chance to air my views on the Irish cat situation. www.irishlakemonsters.com

Sources:

1. MYSTERY CATS OF THE WORLD - KARL P.N. SHUKER (1989 - HALE)
2. THE SMALLER MYSTERY CARNIVORES OF THE WESTCOUNTRY - JONATHAN DOWNES - (1996 CFZ)
3. THE STEP ON THE STAIR - SHEILA ST. CLAIR (1989 – GLENDALE BOOKS)
4. ALIEN ANIMALS - JANET & COLIN BORD (1980 – GRANADA)
5. STRANGE CREATURES FROM TIME & SPACE - JOHN KEEL (1975 - SPHERE BOOKS)
6. LO! - CHARLES FORT (1931 - GOLLANCZ)
7. ANIMALS & MEN ISSUE # 27- EDITED BY JON DOWNES (2002 - CFZ)

APPENDIX:
THE COUNTY ANTRIM CAT-FLAP CONFUSION!
by Neil Arnold

- C F Z   Y E A R B O O K   2 0 0 4 -

And so it began, like many 'beast' stories of before, like those 'werewolf of the moors' mysteries, and other ghastly, inaccurate tall tales concerning frothing, bumbling 'monsters - and that was just the authorities and the so-called 'trackers'!

On Friday 8th August 2003, website Ananova reported, 'PUMA' SPOTTED ON ANTRIM COAST, stating:

"A puma has been spotted on the loose on the north Antrim coast. Police officers and members of the Ulster Society for the Prevention of Cruelty to Animals are searching the area. It was around the seaside town of Portrush and the villages of Portballintrae and Bushmills. Inspector Milne Roundtree of Coleraine police warned the public not to approach the animal.

He said, 'This is a large, wild cat which may have escaped from a private collection. We are currently co-ordinating a search with the USPCA to locate the animal. Anyone who sights the animal in the Portrush, Portballintrae and Bushmills area is advised not to approach it and contact police in Coleraine immediately."

A day later and BBC News reported that the SEARCH FOR PUMA CONTINUES, claiming that several volunteers had been laying bait for the animal which had been seen several times. Apparently, whole chickens were strewn in areas that appeared to 'show' where the 'cat' had been despite at this point no-one actually confirming the identity of the elusive predator. Stephen Philpott, obviously not an expert on large, exotic cats said the animal was either, "…a lynx or a puma escaped from a private collection", despite the fact that for over a century large, wild cats have been sighted across Ireland, mainly due to the fact that licensing laws are extremely slack as opposed to the Dangerous Wild Animals Act introduced in England in 1976. Large cats such as Lynx, Cougar, and black leopard are still, to this day, housed in shoddy, dilapidated conditions across Ireland, and although not escapees, are responsible for some of the populations, it is more likely in Ireland that such generations are offspring spawned from cats let go or from those that have escaped from poor facilities.

Mr. Philpott stated that, "There are at least four of these cats on the loose in Northern Ireland and we have not been very successful in capturing any of them. We are very interested to see what it is and where it has come from so we can stop this trade in exotic pets which is a silly practise".

Already confusion was arising among the press and the authorities with some claiming to know the identity of the animal despite no coverage of actual sightings, and various contradictions in the media. Quite what the authorities were doing to track down the 'animals' is unclear, but the flap had started, despite the fact that no-one had considered that they may simply be dealing with a felid that had been established in the area for many years!

Anyway…..**BIG CAT SPOTTED IN CARRICK AREA**, reported Kim Kelly for the

Belfast Telegraph on August 11th!

"Police and USPCA officers are today continuing to lay bait in areas of remote countryside in an attempt to capture a puma which is on the prowl on the north Antrim coast.

"Officers in Carrickfergus have now joined in the hunt (!!!) after a member of the public reported what he believed to be a lynx or puma near Beltoy Road in the town..."

(Now, just to clear a few blatantly obvious details up regarding the already confused identity of this felid. Those who may not be knowledgeable regarding cat species should be told that the puma is a long-tailed, fawn-beige-darkish tan-coloured felid, that can measure over five feet in length. The Eurasian lynx is a leggy, albeit short-tailed, tufted-eared cat with a mottled coat, and those 'authorities' mentioned should be quite aware of this, otherwise they may have well been chasing shadows!)

To continue: "...*USPCA warden Davy Liggott, who is an expert on big cats, having worked at the former Causeway Safari Park, was called to the scene. He has been using his experience to help police track the animal down and has been leaving out whole chickens as bait in areas of remote farmland where he believes the animal could be hiding."*

The same day BBC NEWS reported that the **WILD CAT REMAINS AT LARGE**, and that the animal in question was now, to quote, "...a young puma"!

Kim Kelly of the Belfast Telegraph was back on the scene on the 12th with a report claiming that paw-prints of the creature had been found at Portrush and that police had been patrolling the beach on the look-out for the 'big cat'.

The tracks, believed to have measured six-inches across were found in a field just outside the resort. A family from Leeke Road had seen the animal on the Sunday and described the large cat as, "...black/brown in colour and the size of an Alsatian dog". Once again though this causes confusion. Whilst a puma silhouetted may appear black, there is no resemblance in colour between the cougar and the black (melanistic) leopard, however, in build there are many comparisons. Across the UK reports of cougar (mountain lion, puma) and black leopard (often called black panthers by the press) always describe animals Labrador-dog size, muscular in the shoulder, and having long tails with an upward curve at the end. Some researchers would argue that black pumas exist in Britain, but as such creatures are extremely rare in their countries of origin, I will not touch on this here. However, both the leopard and puma are agile creatures, able to bound great leaps, and reach over five-feet in length and often standing over two-feet at the shoulder.

The most intriguing aspect of the Kim Kelly report comes in the last paragraph of her article in which she states: *"A big cat has also been seen near Carrickfergus, however it is understood that it is not the same animal as the puma spotted in Portrush".*

Based on press reports, it seems that the animal in question seems to be a puma, but could also be a lynx, but isn't the same animal seen elsewhere, although no other species, by August 12th, had been mentioned. It also seems bizarre that the police believe a cat has been released and that they should hunt it as an emergency, despite the fact that other authorities are admitting that cats are loose elsewhere. In fact, if such flaps were police priority then why were the 'big cats' of Laois, Limerick, or Tyrone treated with the same awareness many years ago? The facts are, when the press creates a press, the public become more aware and alert of 'big cats' in the countryside. Not every report will be genuine, but they must all be taken seriously by the police, and in turn the press blow them up as local beastly yarns for the public to get their teeth into. With such interest generated it is then possible to monitor the movements of a particular cat as members of the public phone in to radio stations and newspapers, and more and more 'hunts' are organised. However, with the attention span of a gnat, much of the public become fed-up by the fact that these cats aren't just going to turn up and so silly season soon hits a dead end, enabling these wonderful animals to skulk back into the woodlands and fields out of the public eye. After such trends, which often bring out the fright brigade of binocular-wielding anoraks, wearing face-paint for supposed camouflage and calling themselves 'beast-watchers', reports then filter to the press very inconsistently, and end up as dribbles, until the next accidental spate of sightings. It happens all the time. Forever.

BBC News went on to report that, "...paw prints found in the north coast area of Northern Ireland were probably made by a member of the mountain lion family", the USPCA has said. (I would love to ask the USPCA what other members of the mountain lion family there are!!)

"...there have been 17 sightings of the animal but the rough terrain has made it difficult to capture." (Two points to note here: 1) maybe if the areas had been monitored instead of the '17 sightings' hardly being given a mention, then surely an area could have been determined regarding the animals prowling ground, and 2) how were these authorities going to capture an animal which they couldn't identify, and certainly couldn't pin-point to any kind of area).

Steve Philpott was obviously on the ball when he stated, " It will pick a spot it is comfortable with, feels safe in and then we'll never see it again. To the best of our ability and as far as we can say it is a member of the mountain lion family, most likely a puma." (Sorry for the sarcasm here, but once again Steve, please tell me about the rest of the mountain lion family!!!).

Philpott also claimed that, "...I don't think we'll ever get it and I can't say it won't pose a threat to anyone but they only come out at night (what about the reports in the day-time Steve!?) but there are three other big cats in the wild in the north, one in the Castlereagh hills, one in Sion Mills and one in the Poyntzpass/Armagh area believed to be a panther."

Okay, another few points to make here. Ask many people in Britain their definition of a 'panther' and they will say 'black panther', as if this is a separate species of felid, when, of course, the 'black panther' is a black leopard. However, across the U.S.A. the 'panther' is another name for the puma, the cat believed to be roaming the Antrim coast. Mr. Philpott speaks of the 'panther' as a black leopard, otherwise he quite clearly would have said puma, but, back to the drama!

Press reports were becoming increasingly confused, with various contradictory and confused statements, such as, "…it's a confused animal, unaware of its surroundings", and, "…it's being seen a lot because it doesn't understand its freedom." Strange statements when you consider the abundance of daytime reports of large, exotic cats across the UK each year. Indeed, there must be many confused cats out there!!

On August 22nd, the BBC reported, "A County Antrim shepherd whose ram is thought to have been killed by a big cat, says he fears for the safety of children in the area. Robert Calvin found the dead, 70-kilo pedigree ram at his farm on Thursday. It had major injuries and bite marks to its side. None of the other sheep in the field appeared to have suffered injuries and the attack was not thought to have been carried out by dogs. Police are investigating the incident."

Unfortunately, as seen in the Welsh incident during early 2003 in which a large cat allegedly ate a farmers dog, police intrusion is not always, if very rarely, of any kind of help. Marksman (with obviously nothing better to do at the time, but who wanted to be seen doing something!) were called to various hot-spots across Wales in order to hunt prowling, dog-eating felids but thankfully to no avail.

Days after the ram mauling, police and USPCA officers were called to the Cookstown area to investigate a partly devoured calf found near woodland. Was this the same animal or quite simply a case of people becoming more alert to signs of other large cats roaming Northern Ireland ?

The Ballymena Times reported, **CAT-CH IT IF YOU CAN!** as their main headline when Council dog-warden Nigel Devine joined the hunt for the cat stalking the Province's North coast. Nigel commented to the paper, "We looked for the animal until 3:00 am but with no success. Turkeys have been left as bait in a four-mile radius from where the cat was first spotted in the hope it will stay in the area because there is a regular food source."

Then, shock horror, Irishnews.com claimed that a black leopard had now jumped on the bandwagon of beasts, believed to be roaming remote ground between Cookstown and Dungannon. Was this a case of members of the public reporting black cats as pumas all along? Were they ever aware that black leopards existed? Probably not, and mainly due to the incorrect information given out by the press.

-CFZ YEARBOOK 2004-

The Electronic Telegraph reported that paw-prints found near Bushmills were …either from a cougar or mountain lion (spot the difference, people!), proving that confusion was boiling over within the flap.

It wasn't long though before the 'beast' had become the hound of the Baskervilles! Kim Kelly reported of the CALL FOR SNIPER TO TRACK BIG CAT as, "…terrified residents today demanded a police marksman is brought in to track and kill the puma (definitely a puma now, then?) blamed for ripping apart an 11-stone ram at Causeway Coast Farm.

Thousands of tourists will flock to the north coast for the bank holiday weekend and locals fear children and visitors could be at risk from the beast.

Experts know the animal, which yesterday morning ripped several pounds of flesh from a sheep's carcass, will be forced to strike again to stay alive (forced!!? With enough rabbits, pheasants, rodents and the occasional lamb to last it a lifetime, I really don't think children are high on the menu!).

By the this time police and other authorities obviously had no clue as to what cat they were dealing with, where it was moving or where it had even been, hence their belief in a marksman who quite simply sits holed up one night on a farm and wastes his time as a cat can smell him a mile off and move on, but hold on….there are developments, some astounding progress!

POLICE STALK BIG CAT screamed the Sky News headline on 24th August. "Police tracking a puma think they have cornered their quarry in a field (obviously long enough for the press to write and broadcast a whole sensational story on it!!). It has been spotted several times in recent days, including a sighting by a woman who said it had been prowling along a road before darting off into some fields. Stephen Philpott, from the USPCA had the job of inspecting the animal's droppings ad confirmed it was an '… exotic big cat' (not a puma or lynx then!!), and that it may be a kind of cross-breed (now this really is a joke!!) and is either a female or a young male (what else could it be!?).

Police say the hunt for the animal has been narrowed down to around six fields and a marksman is on hand, but have called off the search until the next sighting (sorry Mr. Marksman but large cats such as the puma don't live within a territory of six-fields, but hey, you guys said you had it cornered!).

Thursday 28th August, **'BIG CAT' STRIKES AGAIN IN NORTH ANTRIM**, and Chief Inspector Mark Mason believes there may be more than one! Two police officers claimed to have seen the cat and discovered an eaten sheep in the vicinity of the sighting at Benvarden Road near Bushmills. The carcass was discovered at 7:00 am as the officers were driving home from work at the time locals believed the cat was living in Conagher Woods, whilst the Belfast Telegraph were hot on the trail of the 'two' big cats. **NEW FEARS THERE MAY BE A PUMA AND...A PANTHER!** Kim Kelly wrote on the 27th of the month. Allegedly, a photographer snapped the 'beast' on the 26th at Ballinlea Road between Moss-side and Ballycastle. Sammy McMullan from Ballymoney was alerted to the area after reporter Lesley-Anne Henry saw the cat when she was driving home from work at Ballymoney. Ms Henry claimed, "'I've been out looking for the animal, I have been reading and writing endlessly about this puma".

As always, confusion arises, without the press even latching on. Ms. Henry saw a pure black cat, two-and-a half feet at the shoulder with a long tail as it ran out in front of her car. She was adamant it was the 'puma'! She commented, "...it didn't look like how other people described it so maybe I have seen something different. It didn't seem to see me so I got out of my car and followed it but there were lots of bushes and I couldn't see it. I phoned my photographer who was nearby and we drove up and down the road looking for it for about 20 minutes then it ran back across the road the way it had come. Sammy just managed to grab his camera and got a picture of its back legs as it disappeared into the bushes. I was a bit scared but more than anything I was annoyed that I had missed the deadline for my own paper, the Ballymoney Times, that had just been printed."

Days previous a Ballymoney resident claimed she'd seen the cat sleeping in her back garden, but by this time the cat had taken on many forms, ranging from a large, tanned cat, to a small brown cat, to a big, pure black felid.

The photo of the mystery 'cat' was printed on Page 3 of the Belfast Telegraph dated August 27th, and shows an animal disappearing into the undergrowth. However, the animal, which appears to be dark in colour, but not black, resembles a fox, and a bushy tail is evident. Although it is difficult to judge the size of the animal as photographs are often misleading, the photo does not offer any true identity and simply remains another of those blurred entities we have become accustomed to seeing over the years in the press. However, even clear photographs die a quick death in the press.

By this time the 'beast' had been sighted over thirty times, with Mr. Steve Philpott calling it, "...an amateur", in the way it killed, as it was, "...used to having its food on a plate", which is almost a way of conceding that the animal had been released/escaped from a private collection.

Panic gripped the Kilmoyle Primary School at Ballybogey when on the 1st September internet sources spoke of the **SCHOOL ON ALERT AFTER SIGHTINGS OF BIG CAT**, after a sheep carcass was found in a field not far from the 100-pupil strong primary school. The North Eastern Education and Library Board stated that it expected principals and parents to take sensible precautions but that there was no need to consider closing the schools. Principal Caroline Carr commented, "We will liaise with parents and keep the situation under constant review but if we think there is danger to children at all we will keep them in the school at all times. If it hasn't been seen in the area again by Monday we will let them outside but make sure they are supervised at all times."

Things took on a brief, and rather quirky twist around the same time when a Vietnamese pot-bellied pig was blamed for several attacks on livestock!

The Belfast Telegraph churned out one of the most hilarious articles pertaining to the now folkloric beast, when, on 30th August Kim Kelly wrote, under the heading, **WILL KILLER CAT TURN ON HUMANS ?,**

"If all reports are to be believed people are now looking for a light brown, black, large cat, which looks like an Alsatian dog or maybe a fox, is between one and five feet high, and has a bushy tail and staring eyes. The beast can apparently jump five-foot fences, climb trees, swim rivers and has an unrivalled knowledge of the north west, having been spotted at Portrush, Bushmills, Ballymoney, Cushendall and Cookstown"(And this is where werewolves are born from!!)

Army patrols, RAF spotter planes, 24-hour farmer surveillance, including landowner Brian Wotton, who claimed to see a small, sandy-coloured cat, could not flush out the mystery cat which now had become nothing more than a suspected escapee, when the USPCA urged the owner of the cat to, "…get in touch."

Allegedly, police saw the cat one night with night vision equipment but before they could act the animal was disturbed by a car-load of 'cat trackers' attempting to film the felid, and various 'big game' hunters (in other words, trigger-happy nerds with too much deodorant on making too much noise) had also been warned to steer clear of areas where the cat had been sighted to avoid intrusions upon police procedures.

A Catherine Henderson then saw the not so elusive cat on Colyfin Road with her daughter during mid-September. She was driving home from work with her daughter at 5:45 pm when a big black cat ran across the road. Catherine's husband was also in the car, and so they pulled over, got out of the vehicle and looked across the field where the animal had ran. The cat was standing in the field looking at the startled family before it made off. They all described the animal later as being dark grey with a long, curving tail. They phoned the police and two officers turned up but their search was fruitless. They were armed with shotguns and told the family that they would shoot the animal dead if they saw it.

Ananova, 24th September: **TRACKERS CLOSE IN ON BEAST OF BALLYBOGEY.**

"Police say the net is closing in on a big cat running wild on Ulster's coast. Police and animal welfare chiefs said they had narrowed the 150 square mile search down to a wooded area near the village of Ballybogey (even though a few weeks ago they'd narrowed the search down to a few fields somewhere else!!) outside Portrush, Co. Antrim. The 'big cat', thought to be a puma or panther (here we go again!) has been blamed for mauling several sheep after being set loose by a rogue owner who faces prosecution once it is caught."

Under the heading, **HUNT FOR BIG CAT TAKES TO THE AIR**, for the Belfast Telegraph, Kim Kelly wrote that,:

"USPCA experts armed with tranquilliser guns joined PSNI officers in the biggest hunt to date to track the killer puma which has mauled livestock and terrorised the north Antrim farming community (despite the fact it is meant to kill to eat and hadn't terrorised a single person, with most sightings being from vehicles or at a distance!). The puma is thought to have escaped from a private collector who has as yet not come forward to claim ownership of the animal. Today's search is focusing on a ten-mile area between Ballycastle and Coleraine where the wild cat has been spotted on numerous occasions. The USPCA is confident that the beast will be caught today and experts have been monitoring its behaviour patterns in an attempt to capture it before it kills again. It is understood that the last known sighting of the animal was at approximately 7:00 pm on Monday evening when it was seen stalking cattle near Ballybogey. The elusive animal, which has killed four times (only four times! That must be one starving hungry cat!) appears to have been keeping a low profile in recent weeks (which is unusual behaviour for one of the most elusive animals on earth!) and no further killings have been reported to police (but what about all the rabbit murders? This cat is a real serial killer!).

On the 25th of September the U.T.V newsroom reported on the net that, **NI AUTHORITIES RECOVER 16 WILD CATS,** as the Ulster Society for the Prevention of Cruelty to Animals revealed they were looking after six tigers, a lion, and leopard recovered from a bungalow in Omagh during 1997 as well as eight other large cats.

Search teams out in the countryside looking for the 'beasts' of Ballybogey were given briefing notes which outlined possible outcomes and conclusions for the 'beasts', these four being that:

1. The animal could starve to death (despite the thousands of sheep as well as smaller prey around).
2. It could be caught in a trap (although this rarely occurs).
3. It could be tranquillised and then recaptured,(despite the fact no marksman at any point had any kind of cat in their sights.)
4. It could be shot dead by a police marksman...(I doubt!)

Rather bizarrely at this point it seemed that the police really believed they could corner an elusive, agile and adaptable creature that could well have a territory of up to seventy-square miles or more, and often hunts at night! They really believed that by scouring locations for hours, or even days after the last sighting, they could narrow its territory down to just a few fields, but it seemed as though they just wanted to appear as though they were doing something positive when in fact they didn't have a clue. The

BBC website reported that, "…a leading animal charity has said that not one, but two wild cats are roaming north Antrim's coastline. The USPCA said a black panther was living near the village of Ballybogey, outside Portrush, and a brown-coloured puma was roaming the hills near Ballycastle. The claim was made as a dawn to dusk operation by the USPCA members and police officers on Wednesday to catch the wild cat drew to a close. The USPCA said both animals were released near Bushmills in July.

Reported sightings by members of the public showed discrepancies in the colour of the animal seen. However, the charity said it played down the differences, lest the public became concerned about two wild cats on the loose. The USPCA is now withdrawing from efforts to track the cats, saying they will melt away into the background. However, police said they would continue to investigate sightings of both animals"

The Belfast Telegraph reported, **WILD BEASTS 'WERE FREED DELIBERATELY'**. Kim Kelly reported that, *"…the puma and the panther, wearing a silver collar have now made their home between Bushmills and Ballycastle."*

And apparently, *"…a helicopter swooped on the beast and police marksmen stalked across fields with guns after they were alerted by a member of the public. Varying reports from the helicopter crew indicate that the black beast was within their sights, however, another USPCA member said the 'beast' may have been a Labrador dog".*

On the 29th September, the Telegraph reported that hunters, armed with guns had been tracking the cats which had been on the loose for seven weeks. Stephen Philpott commented, "…we are saddened to learn that groups of men with lights and dogs and doubtless guns were seen entering woodland at night and disturbing and damaging the environment."

The newspaper also reported that, "…over the weekend the puma was spotted close to a hen house at a farm near Dervock, whilst other reports indicate that the cats have been spotted roaming housing estates in Bushmills."

However, the best evidence yet to support the theory that a cougar (puma, mountain lion) was on the loose came on the 30th September when the Belfast Telegraph reported, **NEW IRISH BIG CAT PICTURE**, and of the cat photographed that, "…moved like a hunter" and was, "…bigger than a collie".

Ashleigh Wallace reported, "…the dramatic picture which provides the strongest evidence yet that another puma-like cat is on the prowl in Northern Ireland – this time in the heart of the Co. Down countryside." (However, due to lack of consistent reports it is difficult to say whether the Antrim cat was indeed the County Down cat, because, as mentioned previous, large cats have been seen around County Down for many years.)

The cat was snapped by Doctor Brendan O'Donnell who was with his wife when he saw the animal in July 2003 near to Slieve Bearnagh in the Mourne Mountains. The newspaper reported that the animal was, "....a large, black, puma-like cat prowling in a field", however the photograph clearly shows a dark tanned cougar striding through the long-grass. The animal is certainly not a domestic cat. Mr O'Donnell told the paper that,

"...it was unbelievable, it was completely different to a domestic cat in the way it moved, which to me was like a hunter and the long tail was incredibly distinctive."

However, whilst the photo appears to stand as conclusive proof that a puma is at large, the newspaper reported that, *"...this new evidence that a large black cat is roaming close to the Mourne Mountains comes as the two wild animals still at loose in north Antrim continue to evade capture"*, and then to confuse issues more the paper goes on to report that, *"...following the interest stirred by sightings of two big cats currently roaming the north Antrim coast and the sighting of a lynx-like creature in Sion Mills, the USPCA's chief executive Stephen Philpot revealed earlier this year that, '...we have four big cats on the loose in Northern Ireland that we have had regular reports on over the years."*

However, vague reports of 'black pumas', 'lynx-like' cats and other hazy felids certainly does not paint an ideal picture of the established species roaming Ireland's scenic countryside. If the press and the authorities had taken the reports seriously long ago then some kind of outline of territory, population etc, could have been determined. The facts are, nobody can tell if these cats have been recently released, as reports of cats with collars remain inconsistent, and the public as well as the press have yet to decipher the difference in species of wild cat.

As of late Autumn 2003, the press reports of the Antrim 'beast' have died down (in other words, people got bored when the police etc, proved they were not able to out manoeuvre a cat), and in typical fashion most 'researchers' have given up the ghost as such simply because the cat will not come to them. Unfortunately, hunters and researchers are sometimes not all too different, as both parties want their own reward, and whilst the researchers antics may not be harmful to the animals in question, they simply want their name in the newspapers, in the same way hunters want big cats in their trophy cabinets. Farmers around areas such as Ballymoney are still losing sheep to the elusive predators and at Bushmills several other landowners have caught a glimpse of a large feline form. These incidents have not been reported to the authorities.

The Antrim cats may well have been released into the wilds during July 2003, but they could also have been in the area for many years. There are so many cats roaming the United Kingdom and each of these animals may well be offspring from generations that have lurked in the rural shadows for centuries, but flaps are always created when the press jump on the bandwagon, create their local 'beast' tale and all hell breaks loose. Large cats such as black leopard and puma have prowled on to housing estates during late hours for many years, this is nothing new. Marksmen have allegedly come so close to shooting cats from Bodmin to the remote valleys in Wales, but the regurgitations are always the same and so is the confusion. The public becomes more aware during these silly flaps, and whilst this means more cats may be seen, there will also be hysteria, misinterpretation and always someone who likes to get their mug in the newspaper, just so they can call themselves the 'beast hunter' or the 'tracker' when the reality is they are just as dangerous to these animals as the press, the police and hunters. These cats always melt away. Only a minute fraction are ever captured, hit by cars or found dead, and a minute fraction of the public will some day be attacked by a hungry cougar or angry leopard, but that's life and it has come to this through the ignorance and the confusion caused over the years by the sort of people who have played a major part in the drama of this story, and it repeats itself time and time again.

There are many, many large cats at large across Ireland and populations will continue to explode even when the law on keeping such 'pets' becomes tighter. There is nothing anyone can do, and there is no-one out there able to do it. When we do not understand these creatures we call them 'phantom panthers', when we cannot find them we call them a myth and yet there is a variety of different species across the UK inhabiting the remote woodlands, dense forests and overgrown valleys, and there are more than people realise. The main problem with us humans is that we cannot deal with the fact that we are being out-foxed as such by an intelligent creature, a creature able to hunt in pitch darkness, a creature able to climb trees, to leap great heights and to almost disappear from view, which is why across the world such felids are known as 'shadow cats' and 'silver ghosts'. However, it also helps when the people looking for these beautiful creatures are stupid.

- With thanks to the Centre For Fortean Zoology, Nick Sucik and Gary Cunningham.

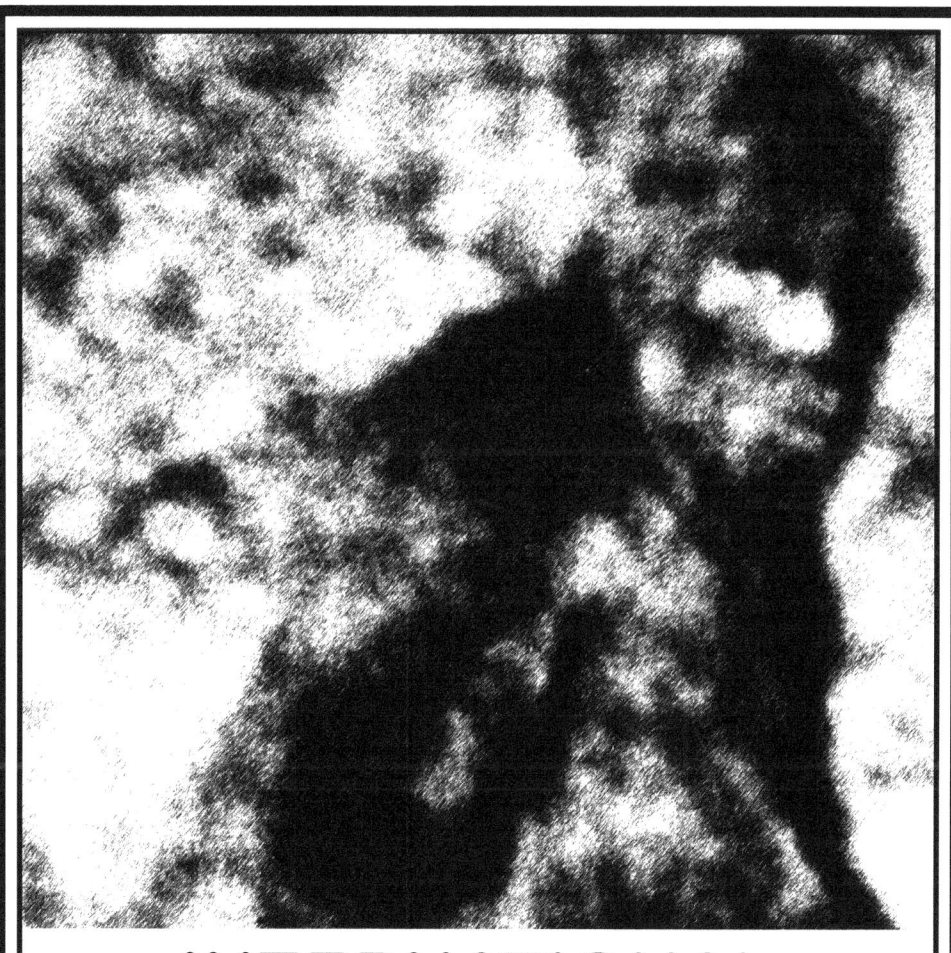

INTERNATIONAL BIGFOOT SYMPOSIUM
WILLOW CREEK, CALIFORNIA
SEPTEMBER 12-14TH 2003.
by Paul Vella

-CFZ YEARBOOK 2004-

In 1972, when I was little more than a toddler, I was sitting on the living room floor of my parent's house in Hertfordshire when the BBC screened a documentary called 'Fabulous Animals', which changed my life forever.

The documentary, which many of you will have seen, showed what is the most studied 58 seconds of film ever made, second only to the Zapruder film, namely the Patterson-Gimlin Film of a 'Bigfoot', filmed on October 20th 1967 in Bluff Creek California.

Over the years, I have read almost every book on the subject, and have corresponded with literally hundreds of people across the world. I have tracked down some of the world's rarest books on the subject, and even have a signed first edition of John Green's seminal work 'Sasquatch: The Apes Among Us'.

So, you can imagine my reaction when a friend from Minnesota told me about the first International Bigfoot Symposium to be held in Willow Creek - after seeking permission from my wife, and promising to do the ironing for the next year, I booked a flight to San Francisco. I called the Bigfoot Motel in Willow Creek, and discovered to my horror that all the rooms were already booked, so I decided on a campervan rental instead.

The date came, and on 10th September this year I landed in San Francisco, sorted out the details of the camper rental and hit the road. Once north of San Francisco, I took the beautiful costal road north, stopping overnight in Bagoda, the small town where Alfred Hitchcock filmed 'The Birds'. The schoolhouse and church are still there, and don't appear to have aged one bit.

The following day I continued my drive north, eventually moving off the highway onto the 'Avenue of the Giants', a 32 mile stretch of road that parallels the highway, and takes the driver right through the middles of 60% of the world's tallest trees, namely the Sequoia sempervirens, otherwise known as the Redwoods - if you haven't seen these magnificent trees, then you need to add them to the list of things to do before you die.

I stopped and met 'Bigfoot' friends from Arizona (Dave) and Washington (Jerry) at the Samoa Cookhouse in Arcata on the coast. This is the last remaining logging cookhouse in the United States, where my friend Dave told us with the story of how he had taken his pickup truck down to the Bluff Creek film site the previous Monday, and got stuck down there after the rain, making the road back up impassable. He camped there for three days, and told us that he was very nervous at the time. After dinner, I headed inland to Willow Creek, arriving there around just before midnight. It was too late to start looking for a campsite, so I parked in the China Flats Museum car park for the night.

Having arrived after dark, I wasn't quite prepared for the sight that greeted me in the morning - Willow Creek was surrounded by steep wooded hills, and looked more like

- CFZ YEARBOOK 2004 -

an alpine village than a sleepy Californian town - the rest of California may have been debating on whether or not to elect Arnold Swarzennegger for Governor, but believe me, no-one here was going to care. The mist was slowly lifting, so I went off to find some more friends who had been lucky enough (or so I thought) to get a motel room.

We chose what was possibly the slowest diner on the planet for breakfast. Actually, 'Sherries' is the only diner in Willow Creek, so it was either that or a doughnut from the gas station.

At the table next to us was sat John Green, Demitri Bayanov, Jimmy Chilcut, Thomas Steenburg, Jim McClarin, Dr. John Bindernagel, Bob Gimlin, Dr. Jeff Muldrum and a couple of other people I didn't recognise. These were the giants of Sasquatch research, and I felt like I was a small child in Disneyland.

Breakfast over, and there was a press conference to take care of back at the museum. The speakers were lined up, and the press received an explanation that we were not crazy folk, but actually people who for scientific reasons believed that there was an as yet unknown ape roaming the forests of North America. I don't know whether the press believed them or not, but at this stage I couldn't care less.

The China Flat Museum in Willow Creek is small - it is named after the 'flat' area of land where the Chinese workers lived (you've got to love Americans for their simplicity). Most of the small museum contains local exhibits, logging photos etc, and then there is the small 'Bigfoot' room, which contains a number of plaster casts, a copy of Dr. Grover Krantz's Gigantopithicus Skull reconstruction, a short video and newspaper cuttings. This particular weekend, the so-called Skookum Cast was also on display - more about that later.

Since the museum was so small, the Symposium organizers had commandeered the use of the local school Gym for the weekend - this being a Friday, we had to wait until after lunch before the talks could start.

We recorded the entire symposium, but in true Fortean fashion, all the audio recordings were unintelligible when we listened to them later. It was an extremely hot weekend, and the organisers had installed a couple of very large (but still ineffective) fans, which was just about the only sounds picked up by the recorders. Fortunately, my friend Jerry from Vancouver, Washington spent the weekend busy with his notepad, and has allowed me to use his notes, which I have largely reproduced here. (Thanks Jerry).

John Green opened the Symposium in the spot originally reserved for Jane Goodall - "Even though I have the same initials as her," Green said, "it is a great misfortune that Miss Goodall was unable to attend as her stature might have caught wide notice."

Dr. Jane Goodall had been scheduled to give the keynote address, but instead was meeting government officials in the Democratic Republic of Congo to discuss habitat

Press conference prior to the Symposium

preservation. Interestingly, she was instead supposed to provide a video address, but that failed to materialise - rumours suggested that her advisors had steered her away from having any involvement in the weekend's proceeding.

Green dedicated the Symposium to the memory of Bob Titmus, noting that Titmus had almost single-handedly started the field of Bigfoot research, bringing to Willow Creek, Ivan T. Sanderson, as well as Green himself. Green told us that many of the items on display in the museum were donated by Bob's wife, after Titmus' death.

Green believes Sasquatch represents a single form of "bi-pedal wildlife" - and made the following points about the Sasquatch phenomena -

1- The story of Sasquatch is one that can't be shaped to fit our fantasies... we must be willing to change our "image" of Sasquatch to fit the facts.

2- We are dealing with an animal... and the study of Sasquatch will require more then one cadaver for dissection purposes.

3- Green believes Sasquatch is an animal, because, not only of what he has learned in his years of study, but because of what science has learned of human origin.

4- There is nothing new about claims of detailed viewing of Sasquatch, or for that matter, people who see, or smell, something every time they go into the woods. Every researcher has his, or her, cross to bear... "mine is the Albert Ostman story." he told us.

5- But there are detailed, as well as seemingly reliable accounts... Glenn Thomas and William Roe, for example. Ruby Creek, which came to light in 1957, produced a casting almost identical to a casting in Bluff Creek, CA.

Green went on to give us a summary of the cases he has personally investigated over the years. Of 3,647 cases in Green's files, only 40 have a male and female together. Smell is mentioned in only 72 reports, with 14 described as "strong," 4 describe a "mild" smell, while 26 reports mention no smell from a distance of 10' or less. Green believes any smell may a "fear odour" as described by Diane Fossey. Green went on to say that 1182 of his sightings were by hunters, loggers, or hikers. The most common sighting is "the one seen on the side of the road."

Green told us that whist it was very tedious entering all the information from his card files into his computer, that it was worthwhile, as it started to show certain trends such as migration, time of day, time of year, moon phase, etc.

You will recall that the death in 2002 of Bigfoot hoaxer Ray Wallace led to numerous newspaper reports that 'Bigfoot was Dead'. On the subject of Ray Wallace, Green said the first paper to print the story reported it in a responsible manner, "but the rest of the press was highly irresponsible". Green went on to say that the Wallace family doesn't even know how the fake footprints were made.

One of the main factors against the Wallace claim is the fact that the discovery of the tracks in the Bluff Creek area strongly interfered with Wallace's contract to construct Bluff Creek Road, costing Ray Wallace money, as he started losing construction workers, causing the construction to run overtime.

Green warned researchers to be wary of associating themselves with people that claimed that they habituated with Bigfoot - it was clear that he was referring to Mary Green's ridiculous book, 50 Years With Bigfoot: Tennessee Chronicles of Coexistence.
Green ended his address by telling the delegates that when it came to Bigfoot research, he was the past, and that this symposium was all about the future. When I spoke to Green later, he said that he never imagined that there would be no resolution to the mystery some thirty years after the publication of his final book. He says he is too old to continue at the pace he has been working, and wants to hand his records over to an organization that will use them properly.

Apelike anatomical and behavioural characteristics of the Sasquatch - Dr. John Bindernagel

Dr. John Bindernagel started his talk by asking "What is it that makes people think they've seen Bigfoot and not an upright bear?"

He summarised the answers as follows:

- Long arms, short neck, flat face
- Female creatures seen
- Short thick neck
- Large nostrils (like a gorilla)
- Pointed (or somewhat pointed) head... "a sagittal crest" - an attachment of muscles to the upper jaw
- Sagittal crests may occur in either male or female gorillas, with size appearing to be the determining factor
- Deep-set eyes (an almost mythological description)
- Pursed lips (seen in NW totems and in pictures of chimps)
- NW totems exist with a thunderbird on top and a Wild Man of the Woods on the bottom
- Wild Giants of the mountains throw rocks at viewers, as do chimpanzees
- Chest-beating (as the gorillas do) are seen in Sasquatch
- The tracks are prime evidence... "tracks often appear as if something was walking on a tightrope"

Dr. Bindernagel did however go on to say that people believe in Bigfoot, or Sasquatch, "because they like to believe in it!"

It should be noted that Dr. Bindernagel's presentations and papers have been rejected by wildlife meetings and seminars "due to recent stories in the press" (apparently referring to the Ray Wallace stories)

Midfoot flexibility, footprints and the evolution of bipedalism: Perspectives on the Patterson-Gimlin film - Dr. Jeffrey Meldrum

Dr. Jeff Meldrum began his talk by telling us that the hallmark of the homonoid locotor system is the grasping toe.

Dr. Meldrum gave a fascinating talk about the Mid-tarsal break in apes and his research into it - used as a test ape was "Jason," walking repeatedly across a box filled with sand, to get an example of the mid-tarsal break. After trying the sand-box test the first time, Jason was put into a holding cage, watching as Meldrum and Jason's handler raked the box smooth.

Following his second attempt walking through the sand-box, Jason looked back at the holding cage, then at Meldrum and his handler (Jason's, not Meldrum's), then back at the holding cage. Jason then went back to the box, and erased his footprints with his hands!

Human footprints show a pressure ridge just behind the toes. Bigfoot tracks seem to indicate the presence of a mid-tarsal break.

Following the Patterson/Gimlin film, Patterson chose to cast tracks which were too "perfect" - so flat that nothing could be learned from them. Lyle Laverty took a series of photos which show a mid-tarsal break (tracks show up very well in Bluff Creek soil.) Bob Titmus cast 10 consecutive tracks, one of which was the same as in the Laverty photo.

Meldrum concludes that "The Patterson film shows flat-footed mid-tarsal breaks and bi-pedalism!"

Meldrum went on to show us a Blue Creek Mountain "half-cast" of a 1967 track which "clearly shows mid-tarsal breaks.", and a Blue Mountains, WA ('96) photo which shows toe slippage and typical front half-track.

Dr Meldrum ended his talk by saying that he believed that "Hominids may have been bi-pedal for over three- and as much as seven-million years."

Dermal ridge evidence in footprint casts - Mr. Jimmy Chilcutt

For me, Jimmy Chilcutt's talk was the most fascinating - many of you already know that I work as a Forensic Examiner and Expert Witness, so for me, it was interesting to see another Expert Witness's 'Evidence'

"My basic job," Chilcutt began, "is to take one fingerprint from a crime scene and match it to one from all the millions on file."

Chilcutt said that he noticed during an investigation, that Cocaine packages are generally wrapped in numerous layers. By carefully unwrapping these packages, Chilcutt was able to isolate prints, on each of the layers, which ultimately led to convictions. The Federal organizations picked up on this very quickly and now Chilcutt works with the FBI, DEA, ATF, and all sorts of "alphabet" organizations.

Chilcutt is a man with a curious mind, and whilst his future research into human fingerprints was going to be unpopular, it is important to us. Chilcutt wondered whether human fingerprints contained any sexual or ethnic characteristics that could be used as an aid in identifying individuals. He was somewhat successful, and states that he can say with an 85% certainty whether a fingerprint is male or female, black or white, but ran into some difficulties as a result of interbreeding of races (it is easy to see how his work could be deemed as being politically incorrect).

As part of this study, he began taking prints from primates (because they don't interbreed.) After being turned down by several zoos, He started with the Yerkes Primate Center, telling inquiring minds he was "investigating the theft of a truckload of bananas!", and found that humans and primates share the same characteristics - arches, loops, and whirls. They are simply present in different configurations.

Chilcutt said that he lived alone at the time, and because he didn't have a wife to nag him, was sat eating his dinner in front of the TV one night with a beer, half-watching the Discovery Channel, when he hears Dr. Jeff Meldrum utter those two words which brought Chilcutt into the Bigfoot field - "Dermal Ridges". Chilcutt called Meldrum the next day, and eventually spent three days examing casts from Meldrum's vast collection.

Chilcutt came away from those three days convinced "...there is an undiscovered North American ape!"

The first cast had dermal ridges, but had been double-tapped, by using human fingerprints to make the toe prints look better. He put that cast aside.
The Walla Walla, WA casts (13-inches long) exhibited "clear ridges with characteristics", with ridges going down the side of the cast.

He went on to explain that human ridges go across the foot and then fade away, and primate ridges run across the foot at an angle...

To Chilcutt's surprise, he found the ridges on the Walla Walla cast ran down the cast on the bottom of the foot, but also on the side of the cast. The ridges on the cast were also "twice the thickness of human dermal ridges."

A Walla Walla cast from 1987 displays the "same texture of ridges as some Northern California casts.', cast some twenty years and hundreds of miles apart.

In addition, he has found dermal ridges on a cast from Georgia... faint ridges, again running down the side of the foot, and explains that the Skookum Cast shows dermal ridges running down the sides of an Achilles Tendon.

Chilcutt said dermal ridges are not present on every cast... but on casts made before 1999, he is convinced that no one could have known the significance of ridges. But casts made after that date could be made knowing all the information Chilcutt has brought to the field of Bigfoot research.

The following day I caught up with Chilcutt in a corridor, and asked whether his involvement in the Bigfoot field had harmed his reputation as an Expert Witness in any way.

"Absolutely not," he said in his soft spoken Texan accent "If a defence attorney were to ask me about it in open court they know they would be given the same presentation I gave yesterday, which would only re-enforce my particular expertise".

Chilcutt and I agreed that one of the most frustrating aspects of Bigfoot research was the poor handling of evidence, and that it needs a good shake up, so that field researchers treat footprints, film, photos etc. as real forensic evidence, and are collected and handled in a forensically sound manner.

I firmly believe that Jimmy Chilcutt is the best thing that has happened to Bigfoot research in years.

We broke for a Samoan smoked salmon dinner, but since one of our party was not keen on fish, we drove back into town and visited a diner. We bumped into some firefighters from the nearby Indian Reservation, who told us that there were more forest fires on Highway 299. A burger and a beer later, and we returned to the school

Doug Hajicek - Patterson-Gimlin footage: Another Look

Those who attended the CFZ Weird Weekend in October will have seen his documentary Sasquatch: Legend Meets Science, which is probably the most even-handed documentary made on the subject to date.

Doug explained that in 1994, while filming a documentary in the sub-arctic, he walked off the path to relieve himself and discovered 17-inch prints on the beach of a lake (with a 42-inch stride, 3-4 inches deep). The tracks went in a straight line, going "over seven foot tall trees" (tracks going on both sides of the trees).

He went back to the float plane pilot (named Wallace... much laughter at this point) and asked him if they could follow the tracks with the plane, upon which Wallace got angry with what he perceived to be a joke and hurled his clipboard. The subject was dropped.

In 1998, Hajicek contacted the Bigfoot Field Research Organization, and began the early work on what would become S:LMS. John Green lent him his first generation copy of the Patterson/Gimlin footage, and Hajicek had the film cleaned up and transferred digitally to a computer.

He quickly determined that there wasn't enough, with the P/G footage alone, to fill an hour documentary.

He spent hours in his basement looking at the digital footage on a High Definition monitor, concentrating mostly on the head of "Patty.", when he began to notice a bulge which appears then disappears, on the right leg of the animal. He later discovered that there is an injury which occurs predominately in women, called an Erectus Femoral Hernia, which is very similar to what can be seen in the P/G footage.

-CFZ YEARBOOK 2004-

*John Green,
Keynote Speaker*

Dr. John Bindernagel

Jimmy Chilcutt

*Dmitri Bayano
Russian Hominologist*

As part of the documentary, he shopped around and found a forensic animator willing to create a "digital skeleton" by "reverse engineering" from the footage. (The first bid came in at $300,000, another was $100,000, but Hajicek, due to budget constraints, was forced to take the lowest bid.)

The result was a skeleton which appeared to walk in a manner completely different than the movement of humans. The knees would bend, with the legs splay outward... then the knees bend in, in a knock-kneed fashion. The Peruvian-gait horse (bred to walk up hills) are the only animal Hajicek has found that exhibit this manner of walking.

His conclusion - "No suit in 1967 (or today) could create this effect."

Other things noticed in the digital P/G footage...

1 - The hair appears to bristle when Patty looks at Patterson and Gimlin.

2 - Patty seems to favour her right leg.

3 - A scientist (who was sure the film was a hoax) agreed to watch the digital footage and was able to identify 40 different muscle groups on the animal. He couldn't give any explanation how that could be hoaxed.

During a question/answer period, he told the audience that, other than the footage in 'Sasquatch: Legend Meets Science', all images of the P/G film came from a third-generation copy of the film. (Both Chris Murphy and Danny Perez maintain the original footage still exists - but neither know exactly where it is!).

When asked about the Ray Wallace claims near the end of the year, Hajicek disclosed that the Discovery Channel pulled all pre-publicity just one week before LMS was originally aired in Jan., 2003. Despite the lack of publicity, the documentary was the first to grow in viewership during the hour broadcast.

DAY TWO

Mayak datat: An archaeological viewpoint of the hairy man pictographs - Ms. Kathy Moskowitz

Moskowitz began her presentation by noting the differences between a pictograph and a petrograph - picto is carved into rock, while petro is painted on rock.

Near Tuleville, CA (in central CA), on the Yokut Indian Tribe's reservation, is a faded pictograph of what may be a family group of Bigfoot. The pictograph was first written about by Mallory in 1889 and again in 1929... this time by Stuart. Both referred to the pictograph as The Hairy Man.

The male The Hairy Man. is depicted as 2.6 meters tall and six feet wide... the female is depicted as 1.8 meters high and 1.2 meters wide... and the baby is depicted as 1.2 meters high and 1 meter wide. The baby is positioned under the female's right hand, while male is to right of the female.

Moskowitz drew a parallel to the Glenn Thomas sighting in Estacada, OR (1967) in which he saw an apparent Bigfoot family searching through a pile of rocks for small rodents. Thomas described the small Bigfoot as trying to stay as far away from the male as possible.

Photo evidence case study - Mr. Alton Higgins

Higgins became interested in the subject of Bigfoot following a fishing trip with his cousin, during which he found tracks and scat.

He started his discussion by noting that evidence can be divided into several categories...

- Testimony - anecdotal evidence
- Signs or indications - physical evidence
- Photographic - this category has "proof" problems; it is neither anecdotal, nor physical

Photographic evidence can be split into numerous categories- 1) obvious fakes, 2) more sophisticated fakes 3) unclear, but possible, and 4) equivocal image evidence

An investigator, as well as a casual person interested in this field, must always question any evidence presented - although it may not be easy to dismiss some images.

Higgins always looks at the 3 "C's".

Context - always explore the context in which the photos were taken. They should not emerge from a void... photos are only one manifestation and should parallel reports and other evidence.

Other evidence may include hair - this is controversial. Samples taken in North. America have been "identified" as primate. Footprints are a very good indicator. Handprints are found very rarely, no dermal ridges have been confirmed on handprint casts, as yet.

Character - includes consistency, reputation, truthfulness, and motive (whether it be financial of fame/prestige)

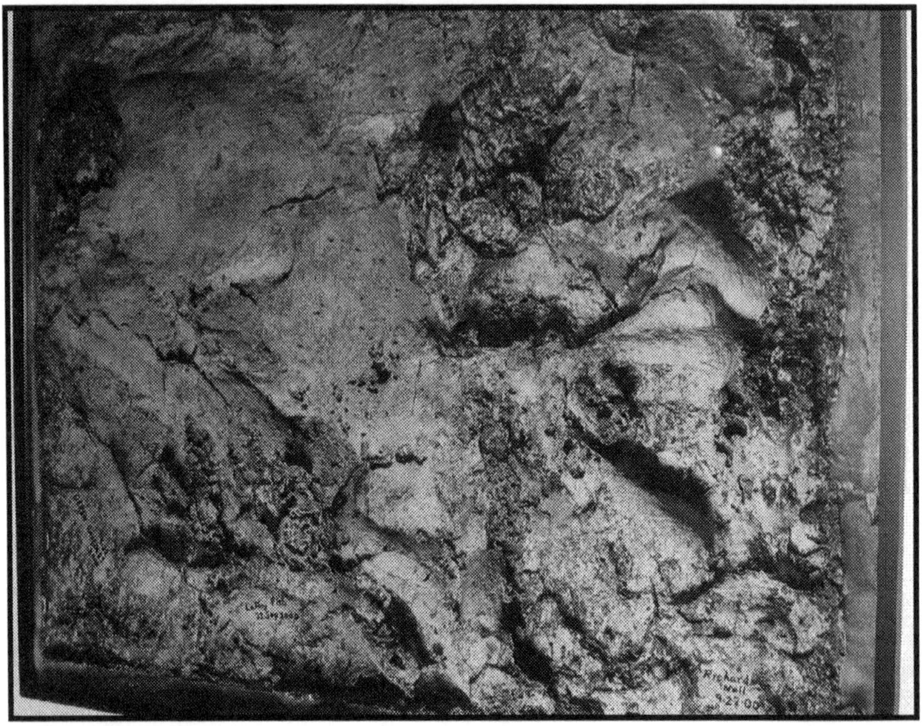

The Skookum Cast

Skookum Cast heel imprint detail

Comparisons - the ability to identify the location of the original, to enable investigators to take measurements and perhaps stage a re-enactment of the photo (to compare height and width).

For so called 'confidentiality' reasons, we were asked not to photograph the photos we were shown, but I have to say that they were not the most convincing photos I have ever seen.

That said, we were also presented with before and after photos which showed that the humanoid shape in the woods was not there before or after, but that is pretty much all I can say on the matter.

In September of 2000, the Bigfoot Field Researchers Organization conducted an expedition in the Gifford Pinchot National Forest of Washington State. This expedition used various techniques and devices to try to lure a Sasquatch near enough to the base camp to establish thermal imaging. Rick will discuss the expedition and its results - a 200 lb. plaster cast of what appears to be the lower torso of a hair-covered primate

Noll gave a short history of sightings reported from the vicinity...

- 1847 - stories told of Mountain Devils

- 1924 - Fred Beck and the famous "Ape Canyon" story

- 1930 - tracks were photographed

- More reports from '56, '64, '65 and '69.

Noll noted that there is a strong belief in Bigfoot's existence in the area, mentioning the Skamania County, WA ordinance banning hunting of Bigfoot and Whatcom County, WA's declaration of a "Sasquatch Hunter-free zone"

The Skookum Expedition was a joint effort of the BFRO, the Discovery Channel, and Animal X (an Australian animal channel.) They used an early example of Thermal Imaging (black and white), as well as pheromone chips and sound blasting, using the "Tahoe Screams" as their calls, and baiting piles.

It has to be said here that I, like many others have seen the Animal X program, which I find interesting because this 'expedition' showed us several people and camera crews driving into the mountains, using these pheromones to waft downwind of them, but it never seems to occur to any of the scientists and researchers that the smell of their pickup trucks, cooking, aftershave and urine will also waft downwind. No one had used the thermal imaging camera before, and they did not have sufficient battery power to use it for more than a few minutes. Despite that, it seems that they got lucky with one of there bait piles, which had been deliberately placed in an area with wet mud.

The assumption was that a Bigfoot would simply walk to the bait pile, leaving footprints to and from, but, the resulting cast appears to show that the Bigfoot sat down and leant across the wettest mud.
Other body impressions have been found in the past, but none have been studied in such detail.
Multiple heel strikes (from two feet) were found, also found elk tracks, as well as coyote, on the top of the impressions. Dermal ridges were found on the "heel."

Noll showed a video of an elk kneeling down on the ground... folding its front legs beneath it, before going to its knees... then repeating the process with its back legs, pointing out that if the impression had been made by an elk, then there would be footprints present under the body impression.

He noted gorillas and chimpanzees both sit down to pick up their food, but I will defer any comment on this to Richard Freeman, since he knows what he is talking about.

During the Q&A session, veteran Bigfoot researcher Danny Perez asked, "Are you 100% sure that the cast represents a Bigfoot?"

Noll simply answered, "No"

At this point, Perez furiously scribbled in his notebook, but did not write down what Rick Noll had to say next.

"That would be stupid," continued Noll, "because none of us saw a Bigfoot make the impression."

Perez: "If not Bigfoot, what else could it be?"

Noll: "I don't know."

He emphasised two points from the start...

   1 - Almasti (Russian name for a Bigfoot-like creature) is real

   2 - Folklore and mythology are both valuable sources of information.

He believes Bigfoot may have migrated from Asia, across the Bering Sea, through Alaska, and from there spread out into North America.

He was unable to attend the Vancouver, BC conference (Conference on Sasquatch and Similiar Phenomena), held in May, 1978 (later written up as 'Manlike Monsters on Trial' by Marjorie Halpin), because the Soviet authorities would not give him permission to travel. This was in fact his first visit to the United States.

- CFZ YEARBOOK 2004 -

Dmitri Bayanov

I have to confess and say that I found Bayanov's accent very difficult to follow - I'm sure his talk was very interesting, but I couldn't make out much except to say that he believes the Patterson/Gimlin film is unique evidence, and that he hadn't seen any three-toed tracks in Russia, but has heard of some four-toed tracks.

The Sasquatch Skin and its Appendages - Dr. W. Henner Fahrenbach

Much of Dr Fahrenbach's presentation was a Primer on Human Skin and frankly, fairly boring. But he did touch on Bigfoot, however, since I fell asleep during this presentation, I am very grateful to my friend Jerry for filling me in on the details later.

Sasquatch Stench: He said most humans begin to smell after one or two days, "... so one can imagine how badly Sasquatch could smell without washing."

Fahrenbach noted only about 105 of the Bigfoot reports contain a mention of a smell. He said this may be an indication Bigfoot may exude the smell when very excited or frightened, much like some male dogs give off a powerful smell from their anal glands.

His former director (at the Oregon Primate Research Centre) believed Bigfoot couldn't possibly be real because the breasts, as seen in the P/G footage, were covered with hair. Fahrenbach felt this position contradicted a book written by the former director, in which he wrote "human breasts are covered with fine hair, as are other primates."

Pioneer Bigfoot investigation Panel - Moderator Mr. Rudy Breuning, Bob Gimlin, Jim McClarin, Al Hodgson, Ed Schillinger and John Green were there in the late 1950s and 1960s capturing film, casting tracks, and investigating the Bigfoot evidence of the times. They will discuss their activities and answer questions about their involvement in Bigfoot investigations.

There was a lot said during this panel discussion, so I will summarise as best I can.

Al Hodgson - "I just happened to be there in '58, '59, and '60... I thought it was a hoax, even after Bob Titmus cast prints." Hodgson said he began to believe when he found footprints on his own.

Bob Gimlin - "I was a non-believer." He met Roger Patterson and got involved with riding around the country simply because he enjoyed riding. Gimlin said Hodgson was the only one, outside of the Patterson and Gimlin families, that knew of the Bluff Creek footage in the beginning. Patterson phoned Hodgeson when they returned to Willow Creek, and said 'Al, I've got a photo of the son-of-a-buck".

Ed Schillinger - was a doctor in the area and heard of a number of encounters from his patients, including a few that came to him with cuts and bruises having run out of the forests having seen a Bigfoot.

Mickey McCarty - in 1972, she had only read of Bigfoot when her 12-year-old son ran into her room, saying he'd heard pigs out of their pen. He'd thrown open the drapes from the window and found himself just inches away from a Bigfoot.

Three days later, her kids were sleeping, in a tent, in the front yard, when a Bigfoot came up to the tent. The kids shined a flashlight on the BF... the Bigfoot ran away, leaving a stench "which lasted for over a month."

She saw a small Bigfoot from about 150 feet. She observed the BF eating berries... when it became aware it was being watched, it "shot me a disgusted look and walked up a pathway up the hill."

Jim McLarin - Bigfoot researcher from the age of 17-26. McLarin is famous for carving the first Bigfoot statue (located at the corner of Highways 96 & 299). He corresponded with Ivan T. Sanderson, Roger Patterson, John Green, and Rene Dahinden.

For him, the "fluidity of movement of the muscles, as well as the light reflecting off the hair" convinced him of the authenticity of the P/G footage. He and two others went to Bluff Creek to view the film site, shortly after the film was shot and he was able to see tracks (in poor condition.) A comparison movie of McLarin, was shot walking along the track of 'Patty.'

Ed Patrick - lived in Hoopa, CA, in 1959. After Ray Wallace reported prints on Bluff Creek Road, Patrick and Bob Titmus went up and looked at prints "17-inches long and 7-inches wide." Patrick said he couldn't make a track on the sandbar - the large tracks were about a half-inch deep and five inches across at the heel.

Tom Slick, the Texas millionaire, became involved with a large-scale expedition (the PNWE), along with John Green and Ivan Marx. They noticed that Bigfoot seemed to go down Bluff Creek around the 1st of November of each year. They set up cameras the next year... you guessed it, no Bigfoot.

They observed, based on the tracks, Bigfoot would crawl under logs rather than hop over them. A helicopter was hired and left gasoline on Oldfield Mountain. Patrick and Marx took a jeep to get the gas and got stuck in the snow. They saw tracks in the snow, coming down the hill, down the road, then turn off the road, and continue down the hill, disappearing out of the snow.

John Green - saw a picture of Jerry Crew holding a cast. Green went down to Willow Creek, with his wife... he saw prints and was very impressed. Bob Titmus mailed Green, who came back to Willow Creek and saw more tracks (covered with leaves). Saw more in '59 and in '63 saw tracks near Hyampom, CA. In 1967, Bud Ryerson called Green, saying, "What you're looking for is here!" over 600 tracks in a row. Green flew down, with a tracking dog and handler. The dog smelled the first track, "and it was an electric shock, the dog's hair on his back stood straight up!" The dog re-

Bob Gimlin

fused to follow the tracks or scent. The tracks were in deep dry dust... then it rained over night, leaving a thin layer of dried-mud on the top. A Forest Service worker drove up to them and said, "I've never seen any Bigfoot tracks in these words." What about these? Green asked.
A Q&A session followed.

The inimitable Danny Perez rose to pose a tough question to Bob Gimlin - "Did you and Patterson stop to shower on the way from Bluff Creek to drop the film for shipment?"

Gimlin - "Danny ..... I can't remember."

Perez - "In all the publicity, surrounding the original article about the footprints, published in the Eureka newspaper, did the editor actually write that article?"

A - No, it was Betty Allen.

Q - "If Jane Goodall came on board the Bigfoot 'field of investigation,' would that increase the likely-hood of more money flowing into the field?"

A - from Green... "It certainly wouldn't hurt." much laughter from Hodgson... "I think the Symposium has been OK with the money we've had." much applause

Q for Bob Gimlin - Do you regret the day you agreed to go with Roger Patterson?

Gimlin - "Not today, not then, but there have been occasions I've regretted it. My wife has taken a lot of ribbing."

And then came the question, from Andy Thomson - a Writer/Director/Producer - that everyone really wanted to ask.

"Mr. Gimlin, I know you have been asked this before, and will be asked this again, but would you mind telling us in your own words would you please recount your experiences on that October day in 1967?"

Gimlin went on to say that he and Roger were riding up Bluff Creek, on the right-hand side. "We came around a downed tree, the creek had re-routed around the tree. As we came around the tree, the horses reacted... to the presence of the creature. Roger's horse reared onto its rear, un-saddling Roger" Patterson it seems was not 'thrown to the ground' as has been reported by many people - in fact, Gimlin went on to explain that Roger had practiced getting the camera out of the saddlebag over and over again, and when he realized why the pony was rearing, grabbed the camera dismounted and ran across the creek and up the sandbar.

Gimlin rode his horse across the creek, and 'Patty' turned and looked at them just as Gimlin dismounted, with his rifle. "I never had an intention of shooting the creature."

Bob Gimlin not only silenced the room while he spoke, but received the only standing ovation of the weekend.

Now, a few words on Daniel (Don't call me Danny) Perez....

I hadn't met Perez before, and if I had met him outside the symposium, I would have assumed that this Jaguar driving Latino that had a moustache that made him look a little like Eddie Murphy was probably a Los Angeles drug dealer.

Perez is the man behind the Bigfootimes (sic) - a newsletter that goes out to around 500 subscribers, and has spent many years researching nothing but the Patterson-Gimlin Film. A few years ago he lost control of his RV driving down to the film site.

Perez is an excitable character, who seemed to associate himself with Lunetta Woods - a woman from Wisconsin, who spent the weekend talking to anyone who would listen about the Sasquatch in her back yard, who could be summoned by watching 'Bigfoot and the Hendersons', and would shape-shift into birds and rabbits to avoid detection !

Lunetta published a book a few years ago called 'Story in the Snow', which you can find on Amazon if you really want to read more. When I read it last year, I had bought it 'blind' from Amazon, and honestly thought it was a children's fiction book !

I only mention her because of John Green's warning about associating yourself with people that claim habitual contact with Bigfoot - I wonder whether Perez listened closely to Green's keynote speech - if he did, I hope he thinks about who he associates himself with.

Christopher Murphy, creator of the scale model of the Bluff Creek filmsite...

Frankly, it was very hard to listen to Mr. Murphy, as they'd added an additional speaker, or turned up the volume. Either way Murphy seemed to be shouting at the audience, or complaining that the overhead projector wouldn't work well enough to show exactly how clear the transparencies he had of the P/G film frames. I'm not sure, but he may have every frame. Either way, his presentation was cut short by Matt Moneymaker, so I am indebted once again to my good friend Jerry, who had seen Murphy's presentation earlier in the year, and filled in the gaps for me.

Murphy met Rene Dahinden in 1993 and used to talk with him three to four times a week. Rene gave him a copy of the famous frame #352 (4"x6") from the Patterson/Gimlin film. He decided to play a joke on Rene and had the photo blown up to 11"x17"... on his next visit he pinned it to Rene's wall, saying "Most men will stare at a photo of Marilyn Monroe, but not you. You'd rather stare at a picture of this!" Expecting Rene to exhibit his "famous" temper, Murphy was surprised to see Rene, instead, just stare at the large photo, finally calling Murphy at home to say, " I can't believe what I can see - bumps and lumps." Rene and Murphy decided to produce a line of posters from photo enlargements produced with "Ceba-chrome" film (producing state-of-the-art enlargements for that time period) that Dahinden had locked in a large safe in his home. But first, Rene had to have a locksmith come and open the safe, having forgotten the combination of the lock.

Murphy said "the experts" maintained, at the time, that it was "impossible to get any detail" in blowups, " because the mathematics say that what you've got is the best you can get." - a 1.2mm high image on the original film frame could only be blown up 85 times before it would become totally distorted - ie, the biggest image would only be 85 mm, or 33.5 inches tall. Yet Murphy, on Saturday displayed an image of 'Patty's' head that was 5.5'-6' tall, with seemingly good quality.

Because of unanswered questions he had about the film... Murphy, using measurements made by Dahinden at Bluff Creek, constructed a scale model on a 2'x2' platform, including all the visible upright trees, as well as the debris seen in front of the Bigfoot, as she crossed in front of Patterson and Gimlin. From Dahinden's measurements and Murphy's scale model; it seems possible to see the film in a different light. For instance, why didn't Patty turn and flee to the rear of the clearing? Because, as the model shows, it was quicker for her to continue towards the trees that she is seen to disappear behind in the film.
Danny Perez, in Bigfoot at Bluff Creek, said no visible tracks were visible in the film, but Murphy thought he could make out at least four! He managed to track down foot-

age from the still missing 2nd reel of film that Patterson shot that afternoon; a BBC documentary, 'Fabulous Animals' (1972 or 3) was the key. From the BBC Murphy obtained 70 frames - which he sent to Yvonne DeClerk (sp), a photo technician in eastern Canada, who was able to piece together the frames into a long strip which Murphy was able to cross check against the suspected footprints in the photos he'd obtained from Dahinden - and he found them to match up.

Other things Murphy could see in the blowups... in frame #61, fur between the toes; frame #307, toes are visible; frame #323, hair on the butt, and to the left of that, the left hand starts to swing from behind the body; frame #350, as 'Patty' turns to look at P/G, chips in the "Ceba-chrome" - not the space-ship or Bigfoot head as others have suggested.

Murphy also pointed out that in the background is not a hill-side, but the beginning of a forest, heavy with underbrush and fallen trees.

He also discussed the question of why the film was shipped to Al DeAtley in Seattle. He said it was his opinion that P/G stayed in Willow Creek to make sure they had captured a Bigfoot on film; in case they hadn't, they would be in a position to return to the area to try again.

Marble Mountain Footage.

We were shown the 'Marble Mountain Footage' that the BFRO had in its possession. Marble Mountain in Northern California is to the north east of Bluff Creek. The video shows a family mucking about outside what looked like a man-made 'hut', before the camera zooms into a figure walking along a ridge of the mountain. It walks for some time before turning 90 degrees towards the camera and walking between several trees.

I have to be honest and say that since the figure was in silhouette, it looked more like a man wearing a coat.

Later that evening, I bumped into Alton Higgins, and complained that the video had been shown without the three 'C's he had spoken of earlier - Context, Character and Comparisons.

Alton told me that he was disappointed that the footage had been shown in the way that it was, but assured me that he had returned to the area to take measurements - the figure was somewhere in the region of 8'6" based on measurements, which puts a completely different spin on things. I wish the footage were available to us, but it shows the family, so they do not want it circulated, and they are entitled to their privacy. A few stills from the movie are available to see on the BFRO.net website though.

The symposium then kind of fizzled out - most delegates had booked a place on the organised trip to the Bluff Creek film site for the Sunday morning, but we had other plans.

All in all, the symposium was simply fantastic - It is highly unlikely that anyone will ever be able to get the likes of John Green, Dmetri Bayanov and Bob Gimlin in the same room again, so it was a once-in-a-lifetime event. This was in fact Bob Gimlin's first trip to Willow Creek since the film was made thirty-six years previously! The event was very relaxed, and all the speakers made themselves available to anyone that had questions – both John Green and Bob Gimlin are now in their 70s, but didn't give any hint of tiring. In fact, Bob Gimlin looked fit enough to take on any man half his age, and positively enjoyed the attention he received that weekend.

When I have a chance to talk to Bob Gimlin earlier in the day, I asked him how he was enjoying the weekend, and he said that it was the first time in 36 years that he really got a kick out of being one of the men responsible for the film, and that he was glad that there were so many people that took the whole issue seriously.

I'm going to give a quick mention here to Lee Murphy, author of a Cryptofiction book called 'Where Legends Roam'. I got talking to Lee in the school car park on the Saturday evening, and he very kindly gave me copies of his two books (Unlike Perez, who gave me a copy of his newsletter and then hung around like a bellboy until I paid him for it).

I've been home now for a couple of weeks, and have had a chance to read Lee's first book, and I have to say it is very enjoyable - the story revolves around a fictional Cryptozoologist, and the first live capture of a Sasquatch. The books are sometimes available through Amazon, and if you get the chance to buy them you should do so.

Adverts over, we went for a very enjoyable meal at the only restaurant in town before turning in for the night.

Sunday morning came, and as usual, I was awake by about 7am - I rolled out of the VW camper, and headed for the gent's toilet at the edge of the car park.

On my way back, I bumped into Al Hodgeson, and got talking to him. I expressed my gratitude at everything he had done to organise the weekend, and he told me that up until Friday morning, he was very nervous - he thought no-one would come and the rest of the town would ridicule him. On the Friday he was really rushed off his feet, and to add to his problems, his dog was rushed to the vet to have an abscess removed. Al was completely knackered by the Sunday morning, but here he was at 7am looking after his museum.

The Museum was not going to open on the Sunday (in fact it is usually only open two or three days a week), Al was in fact waiting for Rick Noll from the BFRO to arrive to take the Skookam cast away, but in the meantime Al very kindly opened up the back door so that I could have a private viewing.

Al Hodgson with the Skookum Cast in the Willow Creek Museum

Al removed the Perspex cover from the cast so that I could take some close-up shots, which was fantastic. One of the things I hadn't noticed on my previous viewing was a boot print - I later found out that this belonged to the late Dr. Leroy Fish.

By the time Rick Noll came to collect the cast, my collection of various friends had started to assemble outside the Bigfoot Motel, and we eventually made our way to breakfast.

We had debated long and hard as to whether we should go on the organised trip, but instead decided that if we all had a few days to spare, then we would go and camp there instead.

So, we loaded up the wagons (actually Bob's pick up truck) with all the essential gear for a couple of days Bigfoot hunting - tents, sleeping bags, maps, torches, matches, burgers, hot dogs, beer - that sort of thing, which took a heck of a lot longer than any of us anticipated, before we filled up the trucks and VW camper with fuel and headed off into the woods.

Now, I had always assumed that Bluff Creek was pretty close to Willow Creek, but we drove miles through two Indian reservations, up mountain roads, before getting off the tarmac and venturing onto the gravel logging roads. Then, we went up and down mountains that were taller than Mt. Snowdon, but didn't even have names, since the mountain next to it always seemed bigger ! This is rugged country - when we got a break in the trees and looked out, all you could see were trees and mountains as far as the eye could see.

Living in Britain as I do, it is easy to underestimate the size of the Pacific North West forests. We were pretty much at the southern end of the forests, which can be followed all the way up through Oregon, Washington, British Columbia and Alaska. Forests bigger than the land mass of Great Britain, with only a few medium sized cities in the way.

I wasn't surprised, then, to hear that the Federal Aviation Administration state that they have lost seventy-five planes in these forests - these aren't planes that have simply crashed - they were planes that had crashed and never been found!

After a couple of hours on the gravel logging roads, we turned onto even worse gravel roads - sometimes stopping to move debris out of the way - these roads don't appear to get moved much !

By the time we arrived in Louse camp, we had driven over 80 miles! Louse Camp is sometimes referred to on maps incorrectly as Louise Camp, but was so-named after the crew building the logging roads suffered a lice infestation.
No such problems for us - we dropped down a steep incline to the camp, and found - well, pretty much nothing!

Bluff Creek

Road from Louse Camp

We sorted ourselves out, erected tents, got a fire lit, and hiked around the area for a while before we cooked some dinner, before sitting up until 2am debating the merits of European beers over American beer - it wasn't a difficult argument let me tell you!

I settled down for the night with the sound of Bluff Creek rushing past the camp, and couldn't help wondering what beasts might visit the camp while we slept.

As it turned out, we had visits from coyote and cougar one of our party was a wildlife biologist who specialises in identifying faecal remains - a very handy person to have with you.

The following day, we piled into three pickup trucks, and headed off for the film site.

The film site is only about seven miles downstream from Louse Camp, but it required another drive of some 40 miles on logging roads to get there. Once we did, we found Steve Iness, who had spent the night at the site in the bed of his pickup truck.
Steve is the son of Orey Iness, a Portland based 'researcher'. I say that with inverted commas, because this father-and-son team are certainly not conventional. They have no interest in capturing a Bigfoot on camera, and when they go into the forests at the

weekends, they rarely even light a camp fire.

Steve told me that they have had a number of close encounters, including one particular night where a large Bigfoot looked down through the skylight of his father's campervan !

Now, I know this is hard for people to believe - we feel we should have photographic evidence, but Steve talked quite naturally about his encounters with Bigfoot, and the fact that he can sometimes go for a year or more without seeing or hearing anything. Neither Steve nor his Father make any money out of Bigfoot - Steve certainly comes across as someone who simply wants to find out more about this elusive creature.

I am a naturally sceptical person, but I came away with two possibilities: 1) Steve is telling the truth - 2) He and his entire family are lunatics. I am inclined to give Steve the benefit of the doubt, since I can understand one person trying to pull the wool over our eyes (ie, Lunetta Woods), but I have trouble believing that an entire family would go to those lengths without at least trying to sell us a book.

The film site itself looks very different to the way it looks in the film. The year prior to the film, a flood washed away nearly all the foliage and a large quantity of earth, sand, shale and trees. In October 1967, the area was covered with a fine layer of grey sand, but today is mostly shale.

It took a little while, but we eventually managed to work out where Patterson had been standing when he first saw 'Patty', and what struck me was that Patty was obviously on the other side of the stream. Where the film goes blurry, Patterson was running across the creek, then climbed some 6-8 feet up the shale bank. It was no wonder the camera shook.

The second thing that struck me was just how close Patterson must have been to the creature, and yet he continued to chase after her.

In a future issue of Animals & Men, I plan to put together a definitive write up on the Patterson/Gimlin Film, so I will not take up any more space than I have to, but for me, it was very useful to see the site with my own eyes. It all started to make sense to me as to why Patty had walked across the sandbar the way she did.

Whilst roaming around the area, we found plenty of signs of other wildlife - I managed to disturb a California King Snake, and found a Banana Slug - a huge yellow slug, which I have no doubt provides something with a tasty meal.

One of our party found a remarkable footprint, approximately 10 inches long with toes, however it was in shale, and not suitable for casting, which was just as well, since we had left the casting material back at Louse Camp !

While we were looking at a pile of scat, and trying to figure out what had left it, we heard a tremendous 'ape scream' - I have heard a number of forest noises, the most frightening of which was a bobcat's growl, but this sounded very different to me, and sent a shiver up my spine.

I walked back to the vehicles in order to send our biologist to look at the scat, and realised that the scream had come from Steve Iness' car stereo! He was playing a tape he had recorded some months earlier in Oregon to members of our party. We listened to it again - it was unlikely to be a pseudo recording, since you could hear a member of Steve's family talking on the tape.

Anyway, to cut a long story short, we headed back to Louse Camp as the light faded - I was sat in the back of one of the pickups, and can honestly say that I have never been so cold in my entire life. Fortunately, Jerry who had gone on ahead had got the most welcome looking fire going by the time we dropped down into the camp, and for that Jerry will always have a special place in our hearts.

We broke camp in the morning, since some of our party had to return to work - due to the remoteness of the area, I was not prepared to stay there without a back-up vehicle, so we joined the convoy of vehicles out of the forest.

I spent the rest of the week in the area, before returning to San Francisco. Just two days after I left the area, there was a daytime sighting reported on Highway 101 where I had been staying.

One thing is for certain - anything could live in those forests. They are largely uninhabited, and, for the most part unexplored. Indeed, many of the local residents make their living by growing marijuana in the forest away from prying eyes !

The forests of Northern California are truly magnificent - if you ever get the chance to visit them, then take it - perhaps you will be lucky enough to meet its most famous resident.

THE CURIOUS CASE OF THE SEAHAM SEA SERPENT
by David Curtis

Every story must start somewhere so I'll start mine here……..

It was only when I gazed across the bar of Winstons to see the eminent fortean investigator, Jonathan Downes, sipping contentedly at his glass of whisky with a full sized inflatable porpoise on his head that I realized my tentative grip on the bow of H. M.S Reality was slipping and that very soon I would be falling into a maelstrom of madness!

In December 2002, I was talking to Jon on the phone and he was telling me that it was all quiet on the fortean front and he dearly wished that something would happen. All he had on his books at that moment was some unusually coloured ferrets that were living in my home county of Durham. Whilst Ginger Durham ferrets would make a quite nice little story it was hardly going to set the fortean world on fire! No, something was going to have to be done and I made it my task to talk to some fisherman friends of mine and see if I could get some sea serpent stories out of their normally tight-lipped gobs.

"Don't worry, Jon," I said. "As my dear departed Nana was fond of saying…something will turn up." Little did I know how prophetic those words would be! Shortly after, in a country park in Northumberland, men and women were coming forward with strange tales of a huge hairy figure wandering around in the woods. The ginger Durham ferrets and sea serpent would have to wait. The game was defiantly afoot! The C.F.Z (a group of investigators who have chased mystery animals across the globe and of which Jon is director) came from their base in Exeter to the north of England to try and get to the bottom of this particular curious case.

It seems that, over the years, quite a few people have seen a strange man-shaped creature in the woods surrounding Bolam Lake. Jon and members of his team came to interview the witnesses and check out the place for themselves. As Jon says himself in his writings on the case, as dusk descended on the woods that night the rooks nesting in the trees that had been making a God-awful racket suddenly became quiet. A member of the investigating team said that there was something moving in the woods. Thinking it may be a deer or a dog, Jon asked the people in the car park overlooking the woods to turn their car headlights on and, lo and behold, there it was! Jon describes it as pitch black, very tall (7 to 8 feet) man-shaped figure running first one way then another through the trees. Back home in Seaham, it all seemed so far away.

The next day I went with the C.F.Z boys to Bolam to do some more investigating but by now it seemed the whole worlds press were waiting to interview Jon and I could see he was going to be busy for the next few hours relating the tale of previous nights shenanigans so I went off on my own to explore. I must admit that it was my first visit to Bolam and I immediately warmed to the place, taking in the sights of ducks, swans and countless other birds flying over this tranquil lake with its picturesque islands and thickly wooded surroundings…. quite beautiful. After a full day of press conferences and investigation it was back to Seaham to toast the success of the expedition or, in other words, get roaring drunk!

Invocatory serpent sculpture on the cliffs at Seaham

Weeks later, the boys had gone back to Exeter and I was sipping on my pint of Guinness in Dempsey's Bar chatting to a fisherman friend of mine named Karl and I brought up the subject of sea serpents once more. 3 stories emerged but none concerning monsters:-

1. A killer whale once used Karl's boat as a back scrubber.

2. A colleague of his caught an elephant! Apparently a circus bound for the port of Hartlepool had an elephant die on board ship so they threw it overboard and he was unfortunate enough to get it in his nets.

3. I have an Elvis impersonating friend named Paul Brace whose father once caught a basking shark in his nets off the coast of Seaham....a very rare occurrence this far north.

All very interesting but Seaham needed a sea serpent and I wasn't getting any nearer finding one.

Just up the coast at Tyneside, Mike Hallowell (a leading light in Northeast fortean and paranormal research) told me of a tale about the dragon cult of Scandinavian sailors who, up until the turn of the 1880's, threw human sacrifices to the mercy of the waves to appease a sea serpent called the Shoney that lived off the coast of Northumberland.

Bugger! The Geordies have the Shoney, the Cornish have a dragon called Morgawr and the Jocks have Nessie. It's not fair!

Something had to turn up and turn up it did! But not at Seaham.

This time, it was down the coast at Skinningrove. A lady angler by the name of Val Fletcher pulled more than she bargained for out of the sea. I quote a newspaper story in full...

FEBRUARY 21, 2003.

FISHY TALE OF DEEP SEA MONSTER

A female angler from Teesside caught more than she bargained for when she went fishing for mackerel, and landed a huge rare species of deep sea fish. Val Fletcher, 40, caught the 11ft 7in-long oarfish monster with a standard rod while night fishing with partner Robert Herrings. The angler, who is 5ft 4in, and weighs eight stone, took 40 minutes to land the fish, which usually prefers lurking deep in the Atlantic. Ms Fletcher, of Marine Terrace, Skinningrove, said: "It was a real struggle to get it in, and then when we did nobody knew what it was - it looked prehistoric.

Val tied the oarfish to a scaffolder's plank outside her house

"Nobody in the village could identify it. We tied it to a scaffold plank and it was outside my house for two days while people came to have a look at it."

The oarfish, or Regalecus glesne (king of herrings), is the ancient mariner's legendary sea fish. They are the longest bony fish in the sea and have a mane-like crest behind a toothless mouth. They can grow up to 30 feet in length and weigh up to a quarter of a ton. Bemused biologists have no idea why the deep water creature, usually found in the Mediterranean and eastern Atlantic, was living in the relatively shallow North Sea. The last one seen in the UK was found in 1981 on a beach at Whitby, North Yorkshire.

Marine expert David Whittaker said: "It is a very remarkable catch, and anyone who finds one should really keep it intact."

The fish, however, has already been cut up into pieces.

The fish known as the King of Herrings was caught with the help of Mr. Herrings! Coincidence? There is no such thing as a coincidence!?

Back at Seaham, my back was really getting up, now even the people of Cleveland had a monster. After reading Jon's book 'The Owlman and Others', I finally learnt how Seaham was going to get its monster... Invocation!

During the 1970's, the Wizard of the western world, Tony Shiels, raised several lake and sea monsters from their lairs by means of invocation. The C.F.Z zoologist, Richard

Freeman, had a stab at this himself in the 1990's.

Every day, as I passed the coast, I focused my mind on imaging a serpent swimming off the coast of Seaham and hoped it would become a reality. This I did whilst making a note of it in the original draft of this article but without telling a soul. Imagine, if you can, my surprise when my dream came true and the monster was sighted by none other than my own daughter and her friends in the summer of 2003!

Terri takes up the story.

'My friends Kayleigh, Hazel, Monna and myself were walking along the cliff tops at the blast beach at Dawdon in Seaham and, looking out to sea, I saw three humps moving in the water just off shore. I pointed them out to Kayleigh and said 'Look Kayleigh, it looks like the Loch Ness Monster!'. She watched for a while and said it could be three seals swimming in single file. I wasn't so sure as they seemed to be moving in unison; sort of one up one down movement as if it were one creature not three separate ones. However, not wanting to be ridiculed, I agreed and we watched until it went out of sight and then continued on our way.'

Seals...hmmm. There is a seal colony about 20 miles to the south at a place called Seal Sands but I've lived in Seaham for 36 years and I have never seen or heard of a group

Terri Curtis (14) showing a typically respectful attitude to her father, reacting to his research

of seals living in the area so I'll reserve judgment. The sceptics might well say I have bribed my daughter and her friends to say these things so all I can do is concentrate my thoughts once again on the Seaham sea serpent and hope that a non relative sees it and spills the beans!

Nothing is what it seems in the strange world of forteana.

In the opening lines of this tale you may remember I mentioned Winston's Bar? Well, it's my bar and it is a converted brick shed in my back garden where my friends and family have spent many an idle hour in the summer drinking the fruits of the vine. Call me mad but it's a harmless madness.

It's called Winston's because it's a Jamaican theme bar done out in the Rasta colours of yellow, red and green and, of course, Bob Marley on the juke box and Marley posters on the wall. But with pride of place going to a framed photograph of Haile Salassie, the last emperor of Ethiopia whom the Rastafarians considered to be God.

Jon looked up at the photo and said with a grin, "you know John Fuller?" "Yes" I replied (John was a recent member of the C.F.Z team) "Well, John is a relative of the Ethiopian royal family and thus a relative to God up there!" You could have knocked me down with a feather! Coincidence? There is no such thing as a coincidence!

Now where are those ginger ferrets?

CRYPTO T.V.

CARTOON CRYPTIDS, ABOMINABLE ADS & MONSTER MOVIES
by Neil Arnold

- CFZ YEARBOOK 2004 -

It was when I saw, back in 1998, an advertisement for women's sanitary pads involving the Loch Ness Monster that I realised cryptozoology, or at least, the monsters that have become household names, yet still a fringe subject, was, in some aspects, becoming a more accessible topic.

Okay, so Nessie, the Yeti, and Bigfoot have become almost fun symbols of a deeply serious subject that involves more than one or two shaky camera shots and ripples in the murky water. Popular, yet poor, television programmes such as 'The X-Files' attempted to bring some of these beasts to the our TV screens, although some forty years previous many rubber serpents and hairy beast-men were invading living rooms across the world, but it wasn't until the 1970s that Charles B. Pierce would combine cryptozoology and television with such effect, with his eerie, lonesome tale "The Legend Of Boggy Creek", a dramatisation of the various Fouke Monster activities which concentrated on the swamp bottoms of Arkansas and its terrified residents.

During the time such a low-budget, drive-in 'horror' was perceived as one of many docu-dramas doing the rounds, but some thirty years later, the images are stronger than ever as more and more film-makers and documentary makers attempt to appeal to and scare the masses with tales of forest dwellers and urban monster myths. Okay, so there's no connection between an atmospheric Sasquatch movie and an advertisement for sanitary towels, but my point is, cryptozoology is indeed making its way into our homes, whether in the form of a picture in a magazine of a pair of knickers sporting a grinning Nessie with the words, "No Evidence" across it, in order to prove that sanitary towels also boast…er…no evidence of….er….stains, or in the form of zany cartoons or big-budget monster trek blockbusters.

Films, adverts, cartoons etc, pertaining to cryptozoological romps and many of the creatures under its wing are common. Whilst they may not be a vital part of the subject, they bring to the masses, the children and a vast audience, information on a more simple, and sometimes more fantastic level. In the CFZ Yearbooks of 1997 (with Michael Playfair) and 1998 I exhaustively listed many films with crypto connections, from over-sized, rampant beasts, to out of place animal tales, and from crypto-treks in search of Snowmen to enchanted tails of water monsters. This updated list is not as exhaustive, merely an update, but one which also includes several cartoons, ads and more recent films to get your teeth into.

Whilst I have studied folklore for many years, researched exotic and out-of-place-felid populations across the world, and exposed the truths behind the Goatmen, Monkey Men and Jersey Devils of this world, compiling lists such as the one you are about to read can be just as engrossing, and hopefully for you the reader it will provide an enlightening and entertaining side to cryptozoology.

The Primitive Man   (1913)
The title says it all really in this creaky, obscure silent title.

Pagan Moon (1931)
A seven-minute Merrie Melodies cartoon rarity set on a South Seas tropical island inhabited by a monster-toothed fish and giant ape.

Superman: Terror On The Midway (1942)
A giant ape called...Gigantic(!), kidnaps Lois Lane in this cartoon short, but thankfully Superman comes to the rescue!

Happy Circus Days (1942)
A hazy cartoon about a little boy and his dog who see a skyscraper-sized ape in a circus!

Haunted Harbour (1944)
A local gang uses sea monster superstition to keep away the gringos! This was a 15-chapter serial from Republic Studios.

The Flying Serpent (1946)
A terrible script, but who cares when a Killer Bird God (Quetzalcoatl) is sent forth from its cage to slay the public!

The Adventures Of Superman: Ghost Wolf (1952)
A 30-minute series (104 episodes). In this episode Lois Lane investigates a timber reserve where many employees have quit their jobs after rumours of a man beast roaming around.

Anak Pontianak (1958)
Malayan oddity, fourth in the series, revolves around a bodiless creature called Polong and a snake-devil, the Hnatu.

Attack Of The Giant Leeches (1959)
I think the title gives this one away!

The Abominable Snow Rabbit (1961)
A groovy 9-minute cartoon revolving around Bugs Bunny and his confrontation with a Yeti.

Dagora, The Space Monster (1963)
A 'globster' rises from the depths of the Pacific and slimes the cities!

Aventura Al Centro De La Tierra (1963)
A low-budget, Mexican movie in which an expedition discovers a new world inhabited

- CFZ YEARBOOK 2004 -

Doctor Who: The Abominable Snowman

by giant spiders, prehistoric survivors, bat-men and a Cyclops!

The Adventures Of Jonny Quest: Werewolf Of Timberland   (1965)
A gold smuggling operation is uncovered in the Canadian backwoods, but the location is prowled by a giant wolf.

Doctor Who: The Abominable Snowman (1967)
The Doctor encounters a walking snowball.

Hillbilly's In A Haunted House (1967)
Starring Joi Lansing, this is not exactly a full-on crypto romp but features a man in a gorilla suit pretending to be Bigfoot!

Creature Of Destruction (1967)
A female sea serpent kills the locals in this dire obscurity.

Yongary, Monster From The Deep (1967)
A sea monster rises from the depths after an atomic blast and heads for Seoul…but doesn't get there!

-CFZ YEARBOOK 2004-

Lokis: Rekopilis Doctora Wittenbacha (1970)
I'm sure you're all familiar with this Polish flick about a mythical black bear that was once a human, that eventually perishes in the snow ? Thought not!

Yeti: Giant Of The 20th Century (1971)
An Italian (!) tale of a 50-foot tall Snow-creature discovered in the wilds of northern Canada.

Schlock (1971)
John Landis directs this movie concerning a rather thin Sasquatch, played by John Chambers, the man allegedly fingered for the Patterson/Gimlin 1968 Bluff Creek Bigfoot.

Blood Waters Of Dr. Z (1972)
An army of swamp creatures, walking catfish to be precise, terrorise Florida.

Chabelo Y Pepito Vs. Los Monstruos   (1973)
A comedy duo take on the lagoon monster!

Sigmund & The Sea Monsters (1973)
A children's series, 29 half-hour episodes. In this episode two brothers befriend a green sea serpent.

The Milpitas Monsters (1975)
A weird creature with bat-like wings and a gas-mask rises from the Bay Area and ends up atop a TV transformer tower.

Six Million Dollar Man: Secret Of Bigfoot (1976)
The Bigfoot doll (boxed) fetches a mint on auction sites across the net, which is more than can be said for the shoddy TV programme.

Bionic Woman: Return Of Bigfoot II   (1976)
The super-woman gets her hands on the hairy thing!

Scooby-Doo:Dynomutt Hour – A Frightened Hound.... (1976)
Scooby and pals look into reports of a sulphurous winged demon that is said to prowl the levels below Seattle.

The Scooby-Doo Dynomutt Show: Who Was That Cat Creature...   (1976)

In New York a cloaked cat-creature commits robberies and the team aim to flush out the culprit!

The Hound Of The Baskervilles (1978)
The Conan Doyle classic gets the Dudley Moore/Peter Cook treatment. Quite atmospheric but as expected, strays from the original. Also stars Spike Milligan, Kenneth Williams, Max Wall etc, etc.

The Incredible Hulk (1979)
The green hero battles the beef of Bigfoot.

The Alien Factor (1979)
Around Perry Hill a creature known locally as the Leemoid, a reptilian freak, is sucking the life from the townsfolk. There is also a tall, hairy man-beast and a clawed, gooey beast called the Inferbyce!

Bigfoot & Wildboy (1979)
A half-hour spin-off from The Krofft Supershow, this episode concerns the missing link who brings up an abandoned boy in the forests of the Pacific Northwest.

Soggy Bog Scooby (1979)
Scooby, Scrappy and Shaggy are fishing in a murky swamp when they encounter the local gill-man!

Scooby Doo Where Are You! – Who's Afraid Of... (1979)
There may be sheep smuggling going on in an old mill, but what about the sightings of a ghostly green man-beast ?

Goofy Movie (1970's/80's)
Bigfoot attacks Goofy and his son Max whilst they are out camping, and traps them in their car. Bigfoot then puts on an entertainment show.

Great Alligator (1981)
An Italian oddity about....yep, you guessed it...an over-sized Alligator.

Island Claws (1982)
A huge crab-like killer critter roams the bowels of Florida's swamplands.

Creepshow (1982)
A decent five-story anthology taken from Stephen King's script. In 'The Crate' a hairy, troll-like meanie is released from its confinement to feed on human flesh.

Mystery On Monster Island (1982)
A Spanish adventure comedy are shipwrecked on an island where they come under attack from huge death worms, rubber dinosaurs, seaweed critters and lizard men.

The creature from the story 'The Crate', Creepshow (1982)

King Dong (1984)
Okay, can't really go into too much detail here except to say, it's pornographic!

Crocodile   (1986)
I personally think the title gives a slight inkling as to what this flick is about!

Demon Of Paradise   (1987)
A Hawaiian lagoon is inhabited by Akusa, the rubber gill-man. And he's not very happy!

Hobgoblins (1987)
Bizarre little creatures escape from the dusty vaults of the city to terrorise the local folk.

Creepshow 2 (1987)
Another reasonable anthology of three tales. 'The Raft' sees a group of youths picked off one-by one by a lake beast that looks like an oil slick.

Dark Age   (1988)
Aussie shocker that tells the tale of the 25-feet long Numunwari, a salt-water monster croc feared by the local natives. However, the locals want this beast protected as it's meant to be the last of its kind!

- CFZ YEARBOOK 2004 -

Demonwarp (1988)
In Demonwood (sounds like a great place!) a bunch of daft youths are picked off by Bigfoot, and it just so happens that some damn UFO probably dropped the Sasquatch off!

Lords Of The Deep   (1989)
Set in 2020, a creature that looks like a swimming bat invades an underwater lab station.

Criminal Act (1989)
News investigators look into reports of humanoid rat creatures dwelling in the city's sewer systems.

Pizza Advert (Late '80s).
A Sasquatch steals the new, much sought-after Pizza and carries it off into the darkness, leaving the delivery guy a bit shocked!

Elves (1990)
Neo-Nazis seeking shelter in Colorado Springs plan to breed an elf with a virgin to create a master race! What!!?

The Crawlers (1990)
Snake-like monster tentacles, the product of nuclear waste, grow from trees and pick of the locals.

Chipmunks At The Movies (1990)
Bloody annoying little things, which, in this short-lived series, find a giant ape, similar to King Kong, and use it for a broadway play.

Capital One Advert   (1990's)
The Yeti stars!

Chips Ahoy Advert   (90's)
With Bigfoot as the star!

Frostbiter: Wrath Of The Wendigo (1991)
Troma movie concerning the ancient snowbeast legend. A spooky skeletal creature with antlers that butchers all the hunters in its wake. Low-budget trash with moments of gore.

Dangermouse: Bigfoot Falls   (1991)
Can't remember this obscurity, but the cryptid falls into this somewhere.

Deadly Eyes (1992)
London becomes over-powered by massive rodents the size of dachshunds. Oh, there

are dogs beneath those skins!

The Simpson's Halloween Special II (1992)
Grandpa tells the spooky tale of 'King Homer', about a 50-ft tall prehistoric ape captured and returned to civilisation, where in turn it invades Springfield.

Humanoids From Atlantis (1992)
Shot-on-Video release about a green gill-man from Atlantis that goes rampant in a coastal town.

Batman: On Leather Wings (1992)
A bat-like humanoid attacks a security guard in this animated episode and Batman is blamed. However, the masked hero soon discovers it's a creature named Man-Bat!

Skeeter (1993)
A take of the '50s sci-fi B-movies, with this, a tale of overly-large flesh-eating mosquito's, but this actually has several sub-plots making it a half-intelligent monster mash.

Tales From The Cryptkeeper (1993)
'Pleasant Screams' is a cauldron of swamp blobs, hydras and feathered monsters.

Are You Afraid Of The Dark? (1993)
In 'The Tale Of The Dark Dragon' the Bunnyman legend comes to life when a group of teens sit around a campfire and tell eerie tales of the monster bunny.

Dinosaur Island (1994)
A campy comedy-adventure involving a team of soldiers stranded on an island full of prehistoric survivors…oh, and scantily clad women in animal bikinis.

Stanley's Dragon (1994)
British TV movie about a nerd who finds a prehistoric egg, from which a dragon is born. Of course, things get out of hand when the scaled beast grows to a rather large size and breathes fire.

Secret Of Roan Inish (1994)
Irish mythological tale set in 1946 about a girl who hears stories of a mermaid-type creature that looks after souls lost at sea. Eventually the girl witnesses the seal-woman but has trouble convincing others the creature exists.

Grim (1995)
Filmed in Clearwell Caves, Gloucester, this film concerns a Neanderthal gorilla monster that does its fair share of growling. Explorers are picked off in the dark depths by the Bigfoot-like beastie.

- CFZ YEARBOOK 2004 -

Magic In The Water (1995)
A better-than-average U.S. film set in British Columbia, with various sub-plots but more interestingly revolving around the local lake monster, known as Orky that is constantly harassed by toxic waste, Japanese news crews and mean businessmen.

The Last Broadcast   (1996)
A creepy, shaky low-budget flick that only came to light after 'Blair Witch...' mania, despite the fact it preceded it by several years. 'The Last....' concerns a team of investigators and local researchers who travel to the New Jersey Pine Barrens with the hope of flushing out the Jersey Devil, but things meet a grisly end in this cool little spooker that combines documentary-style film-making with excellent scares considering its low-budget of $900.

Legend Of Gator Face (1996)
A 'cry wolf' story set in the Everglades about two nasty little kids who make up a story about a swamp beast-man, only for the real thing to emerge later on in the film.

Big Bad Beetleborgs (1996)
In the episode 'Something Fishy' a gill-man creature emerges from the local lake in this children's TV series of some fifty-two episodes.

Gogs (1997)
The primitive, animated family come face to face with a group of Abominable Snowmen in this enjoyable children's series.

The Phoenix & The Magic Carpet (1997)
Not exactly crytpo, but the legendary firebird emerges from an egg in this children's drama.

Father Ted (1997)
In this hilarious Irish TV show, rumours abound about the 'beast of Craggy Island'!

Fruitties   (1997)
These were the characters kids used to put on the ends of their pencils! Anyway, here the gang confront a huge, hairy snow beast.

DNA (1997)
A bit like 'Predator' in the sense a huge, prehistoric/supernatural creature roams the forest bumping off explorers etc.

Ghost & The Darkness (1997)
Based on real-life events, Michael Douglas stars with Val Kilmer, in this tale set in 1896 in East Africa, where two man-eating lions are on the rampage killing villagers, whilst construction on a bridge is taking place. The odd thing is, these lions appear to have supernatural qualities.

- CFZ YEARBOOK 2004 -

Premium Bonds Advert (1997)
All I know is that Nessie starred in this commercial!

Pterodactyl Woman From Beverley Hills (1997)
Just when you thought it was safe to go out....it is!

Ford Advert (1998)
Nessie gets in on the act again as the new Ford at the time purrs along a mountain road and a set of wet humps break the surface of an imaginary loch.

Citroen Advert (1998)
…..and then Citroen cash in too!

Chucklevision (1998)
The legendary Chuckle Brothers meet Bigfoot…it's a shame it never mutilated them.

Detectives
English comedy series in which bumbling detectives investigate reports of an out-of-place bear.

Plenilunio (1998)
A straight-to-video release from Brazil concerning a group of teens working at a local TV station who begin to investigate grisly murders committed by a furry, white man-beast who then begins to hunt the teens.

Gargantua (1998)
A TV movie ran on Fox Network, where explorers find a nine-foot long salamander/toad/lizard mutation…and then bump into its mother.

Godzilla   (1998)
The big-budget Hollywood remake. It's rubbish.

Lake Placid (1999)
Don't go in the water because there's another one of those over-sized alligator things in there…..again.

Evil Beneath Loch Ness (1999)
A low-budget release that turns Ted Danson's 'Nessie' movie into something a little more sinister…yet still as poor.

Mighty Joe Young   (1999)
Remake these films as much as you like but they're still not a patch on the originals.

The Sasquatch Hunters (1999)
Another of those 'mockumentary'-type fiasco's that attempts to create home-made ter-

ror in the backwoods. And fails.

Ice Break Coffee Advert (1990's)
Australian television offers up a drink advert with a difference, as a couple drinking outdoors have their beverages stolen by an orange-brown Sasquatch. They then pursue the man-beast through the woods. The Patterson footage is also used in this ad.

Kokanee Beer Commercial (1990's (?))
Canadian beer, Kokanee, used Bigfoot researcher Rene Dahinden to promote their product. Rene is interviewed about the wonders of beer when a Bigfoot creeps up, steals the beer and makes off with the crate.

Wendigo   (2000)
Too many sub-plots in a frosty tale that strays too much from the Native American legend of the demon of the woods.

Brotherhood Of The Wolf (2001)
A lavish and stylish monster/action adaptation of the Gevaudan 'beast' scare that took place in France around 1765. The beast is well portrayed, characters are strong and the setting makes for a gothic epic of swash-buckling grandeur and monstrous mayhem. The DVD comes with a documentary behind the legend.

Alien Big Cats (2001)
An Australian fiasco revolving around the so-called mystery big cat situation.

Mobile Phone Advert (2000?)
Filmed in the Australian Alps (!!!), snowboarders are pursued by a group of Yowie-like monsters.

Down To Earth (2001)
Pauline Quirk stars in this rural drama series for UK television. In one particular episode, something black and feline is bumping off the livestock. Although we never get to fully solve the mystery the end does provide us with a glimpse of a small black leopard in the fields.

Incredible Adventures Of Hank Bolax (2002?)

Terrible low-budget shorts in which some strange guy encounters a Sasquatch.

The Mothman Prophecies (2002)
Oh dear! This is churned out like an 'X-Files' spin-off, as Richard Gere is plagued by the weird warnings and strange goings on surrounding Point Pleasant in West Virginia. Director Mark Pellington wanted this big-budget adaptation of John Keel's great book to be a fear-based project, but we never get to see Mothman at all. A great opportunity missed here!

Sasquatch (2002)
Not big-budget but another of those docu-drama, stalk 'n' slash flicks that is completely eclipsed by the great Bigfoot films of the 1970s.

Dog Soldiers (2002)
Okay, so it may be a werewolf movie (there's a mention of the 'beast' of Bodmin at the beginning!) but it deserves its place as a classy, darkly humoured British horror revolving around a group of soldiers being picked off in the Scottish Highlands by some nasty hairballs.

Monster Munch Advert (2003)
Good ol' Nessie gets in on the act again, albeit as a rubber suited Plesiosaur.
Post Office Advert   (2003)

The recent, 'I saw this and I thought of you...' ad's reached a crytpo climax when a guy sitting in a tent in the wilderness is sent a Bigfoot book by his good ol' mum with a note saying, 'I saw this and I thought of you'. And then Bigfoot walks by outside!

Pop-Tarts Advert   (2003)
A Yeti attempts to tell two children about the wonders of freezing Pop-Tarts before eating the snacks. However, the kids scream and the abominable creature takes off into the woods with an obese waddle.

Yu-Gi-Oh (2003)
Japanese anime I believe! There's mention of the Chupacabra in some weird card game akin to Top Trumps. Yugi (!) throws in the Chupa card which nothing can beat, and the 'goatsucker' is soon eating Scooby-Doo who finds himself in the stomach of the Moca 'vampire'!

Jackie Chan Animated Episodes (2003)
A whole episode from this series devoted to El Chupacabra.

Will & Grace (2003)
Brief mention of the Chupacabras in this American sitcom.

Jackass (2003)
Brief animated piece from this wacky stunt series showing a group skiing in the mountains where they are brutally mutilated by a Yeti.

Beachcombers: Sasquatch Walks By Night (no date)
Unsure of what this is!

In my 1998 article on cryptids in the movies, I promised that such an exhaustive study would not be updated. However, due to more obscure findings, mainly with regards to television advertisements, and the fact that there were no entries after the year date 1996, this compilation has built into another lengthy list of updates.

Much of the content though would not have been made possible without authors Stephen Jones and John Stanley whose 'Creature Feature' books have provided much entertainment over the years. I do not believe another list of this type will come along ever again, but with more and more real monsters making their way onto our screens, who knows?

THE CUSWORTH CAT
by Dave Baker

- CFZ YEARBOOK 2004 -

For some time now, I have been a member of, and local representative for, the Centre for Fortean Zoology. Last summer, I missed joining with "The Hunt For The Cannock Croc". I was determined not to miss the next opportunity to tackle a Cryptid critter.

My chance came out of the blue, as these cases often do.

An Alien Big Cat report landed on my very doorstep. An account distributed via the Forteana newsgroup, stated a Big Cat had been seen in South Yorkshire a few days ago.

On the 9th August, 2003, 9-year old Charlotte Clarke was returning from a visit to Cusworth Hall Museum, in the countryside just outside Doncaster. Charlotte and her Cousin were passengers in their Grandmother's car.

They were driving slowly along Cusworth Lane; our young heroine glimpsed something strange in the adjacent field. Initially she saw a long tail emerging from tall grass, then, the rest of the beast, crouching and ready to pounce. Charlotte shouted out to her cousin and grandmother, but they did not see the animal. On arriving home, she blurted out the tale to her mother, Diane who was convinced her excited daughter was not lying.

Charlotte poured over a large book of mammals, with good quality line drawing of various felids. She was able to recognise her cat as no less a creature than a Jaguar (Panthera onca), the magnificent, beautiful, South American, animal that has been known to kill large bulls.

Charlotte's story appeared in a number of newspapers, including the prestigious Yorkshire Post, and formed a typically tongue-in-cheek segment on a regional news show, BBC Look North, on Wednesday 20 August.

Immediately after downloading the Forteana report, I received a message from fellow CFZ and YUFOS member Mark Martin, who had also seen the story, and we teamed up to investigate the case.

I called the South Yorkshire Police, and spoke to the extremely helpful and friendly PC Trevor Suter. He gave me a run-down of the story, more or less as I already knew it, and confirmed there had been a number of similar big cat reports over the years from around Doncaster. He explained that the police only investigate such cases if the sighting is recent, as a matter of public safety. If there is no immediate threat to the public, the report is passed to other departments, even those outside the police force, such as the RSPCA.

More importantly, the officer was able to pass on Diane Clarke's mobile telephone number, as she had made it clear she was happy to talk to anyone with a genuine interest in her daughter's sighting.

Moments later, I spoke to Diane, who elaborated on Charlotte's encounter, and said that she had more than just a motherly interest, as she herself had seen a "black panther" two years previously. Whilst talking it became clear that Diane was well aware of the big cat stories circulating Doncaster and it seemed she could be an interesting interviewee herself.

Diane was happy for us to visit and speak to her and Charlotte. Mark and I had already decided to visit Doncaster and investigate the sighting location and the area surrounding Cusworth Hall .We decided to do this and visit the Clarke family on Thursday, 21st September.

CUSWORTH HALL

As we had plenty of time to kill before our arranged arrival time at the Clarke's, we stopped off at Cusworth Hall first. Before exploring the Hall's grounds, we parked alongside the field we thought to be the location of Charlotte's jaguar sighting and examined the field as best we could.

Unfortunately, there was little to see. Since Charlotte's experience the crop had been harvested, and the field ploughed, the long stretch of hot dry weather the British Isles had enjoyed the past couple of weeks precluded any chance of finding tracks. We filmed and photographed the field for the record, and decided to move onto the grounds of the Hall itself. The Hall is actually a museum, and caters particularly for parties of schoolchildren, encouraging them to take an interest in nature and conservation.

The beautiful grounds are well suited to a large feline, with wooded areas, open fields, and some distance from the buildings themselves, a large lake, but again we found no traces of a large animal. It should be remembered though that Cusworth Hall and its grounds are extremely busy, and populated by many visitors and staff; certainly more than enough human activity to keep a wary big cat away, at least during daylight.

We enquired at the Hall if anyone had seen anything unusual in the area. A quick chat with the genial - but decidedly cautious - guide at the door revealed nothing; the man had only heard a vague story about the Jaguar, knew nothing at all about other reports, and suggested we talk to Colin Howes, the Keeper of Environmental Records, who was "somewhere around, probably down by the lake".

By now though, time was catching up with us. We didn't know exactly how to find the Clarke's house and were reluctant to be late. Thus, after making a mental note to contact Colin Howes at a later date, we left the Hall behind and headed for Doncaster, and the Clarke household.

Finding our way to a house in the middle of a town proved more difficult than locating a Museum in the back of beyond. This wasn't helped by Mark driving the wrong way up the motorway, away from Doncaster, because of being too busy yakking about Mothman, The Loch Ness Monster and the CFZ's upcoming Weird Weekend. Nevertheless, once we had realised our mistake, we got back on track, and pulled into a Supermarket car-park to consult a map. Eventually we managed to filter back out into the traffic, and locate the Clarke's home; still with a little time to spare.

INTRODUCING THE CLARKES

We were warmly welcomed by Diane, Charlotte, her brother Reece, as well as the baby of the family. Diane had told me that Charlotte had appeared on the BBC's Look North programme the night before. As Mark and I had missed the report, Diane played us a recording of it.

Typically, it was a fine piece of balanced, intelligent TV journalism. Charlotte was interviewed briefly, a piece of footage was shown - surprisingly enough, of a jaguar - (so someone in the Beeb's research department did their job) and then the journalist-in-the (literal) -field, went out to investigate the location. He wore a jungle-style floppy hat, had a tin of cat food, and made jokes about saucers of milk. Laugh? I thought I'd never start.

After watching the TV segment, we headed into the back garden to film the interview with young Charlotte. She is a very intelligent, friendly girl, and appeared to be an excellent witness. She thought carefully about questions, was precise in her answers, and did not embellish details if she was not sure. Dr. David Clarke [no relation] who later watched the video at a YUFOS meeting declared Charlotte a more competent witness than most adults he had interviewed during his years of research.

Here is part of the transcript of the interview with Charlotte conducted by Dave Baker and Mark Martin, both of the Yorkshire UFO Society and the Centre for Fortean Zoology on Thursday, 21 August 2003. The initial interview was videotaped outside in the Clarke's garden with further questions recorded on cassette, inside the house.

THE INTERVIEW

DAVE BAKER: So, Charlotte, can you tell us exactly what happened?

CHARLOTTE CLARKE: Well, I was coming home from Cusworth Hall, and I saw a jaguar. When I went in the car, I looked in that field, and I saw this thing standing up, or getting ready to pounce at something. So I told my Grandma and my cousin that I saw something, but I didn't know what it was. So I asked them what has black circles and two little black circles, and (was) a sandy colour.

DB: And you stopped seeing this because the car drove past it? Or did the animal go out of sight?

CC: Erm….I think it was still standing in the middle…

DB: But the car drove past so you couldn't see it anymore?

CC: 'cos I was looking…still looking out the back of the car to see if I could see it.

DB: And was it in long grass?

CC: Yes.

DB: So how tall do you think it was?

CC: Erm…about up to my waist…do you want me to…? (Charlotte then stood up and indicated with her hand at waist-level) Up to there.

DB: Right, and could you tell how long it was?

CC: No, not really…

DB: No, well it is difficult to tell sizes with things like that. And what was the animal doing?

CC: Erm…I think it was catching something….like, trying to catch a rabbit, or… something like that.

DB: So as soon as you got home, what did you do?

CC: I asked my mum what has the description I gave to my Grandma, and I looked in a mammal book and saw what it was. And it was a jaguar.
Charlotte then brought the book over and flipped through the pages until she found a double-page spread of big cat paintings. The pictures included a tiger, leopard, cheetah and cougar, among others.

DB: Right so you looked into this book and you were able to find and identify the animal that you saw? And which one was it?

CC: That one (points unerringly at the jaguar) At first I thought it was a leopard, but it didn't have the spots in the middle.

DB: That's right, yeah. And what did you do when you had identified it, did you report it to anyone?

(Charlotte was a bit confused at this point, and could not remember whom her mum had called and in what order, but knew that the police had been called, among others.)

MARK MARTIN: And how far away from it do you think you were, Charlotte?

CC: Ermmmm....about where my mum's standing...not really far... (This was about 15-20 feet)

MM: Not really far? So if I walk down your garden...just down here...and you say I was the Big Cat...... I don't look like a Big Cat but we'll pretend...how far away...if I walk back like this...? (Charlotte stopped Mark approx 20 feet away from her)

CC: Yeah, yeah about there.

MM: About there?...From you...which is about.. less than 20 feet...

DB: And where would you say it's back was...where would it reach on Mark?

MM: So if it was standing here? How far up my legs would it go?

CC: About...about in the middle, where your hand is...

MM: About there?

CC: a bit…up… (This was about 3 foot)

DB: And was it side-on to you, or was it facing you?

CC: Erm…it wasn't even…it was looking at something else…it wasn't looking at me, it was like, erm… like looking at something, like its prey or something…

DB: Right okay, but was it sideways like, moving across, or was it towards you?

CC: Sideways…

DB: Okay. Mark, get on your hands and knees (Mark got down on all fours on the lawn, much to the amusement of Charlotte and Diane, his head facing to Charlotte's right)

MM: Was it like that or…..?

CC: Facing the other way.

DB: Go on, mucky your knees as well… (Mark scrambled around to face the other way.

All joking aside, this simple correction by Charlotte proved to me more than anything that she really did see something. An adult probably wouldn't even have made any correction, assuming that this was irrelevant.)

DB: Okay so would you say it was shorter than that across, or longer?

CC: A bit longer.

DB: So a bit longer, right…Okay, you can get up now, Mark. That's just to try to work out the sort of size it was, from a distance, because its difficult to remember if you don't have anything to compare it to. (From this we can estimate that the animal was about 5ft long, not including its tail)
(Mark walked back to us)

MM: Well that distance there…..20 feet maximum. 18, 20 feet, that's all.

DB: How long would you say you saw it for?

CC: Erm…..a few seconds….about...30 seconds.

DB: So it's a fair time then? (My thoughts are that it was probably not this long, although Diane admits that her mother does drive "very slowly", and Charlotte had said that she had continued to watch the animal through the rear window of the car as they drove on)

DB: Did it have a long tail or a short tail?

CC: Kind of a long tail…'cos people were asking me if it was a cub, but it wasn't. It was too big.

We talked for quite some time with Diane and Charlotte after the interview, which was taped with their consent. Diane has quite an interest in big cats and wolves, and shared a number of additional titbits of information with us. At the time of writing we can't be sure if these are local gossip, urban legend, the literal truth, or variations on these, but I shall look into these further.

These include:

- Sightings of a puma in the Bessacar [the part of Doncaster the Clarkes live] area 2 years previously. Diane claimed she witnessed a "black panther" herself, it walked casually and serenely beside the motorway. Indeed, the so-called "Beast of Bessacar" was witnessed on a number of occasions in 1999 in and around Doncaster, and features in a number of articles in the local press.

- A pregnant panther was released or escaped from a circus which passed through Bessacar 6 years ago. Diane says that the circus also had a male panther, which "pined" without the female and was released by the circus on its way back from the region. Diane says that this is on record in the Doncaster Public Library, but although I have not yet had the time to check this out, it is true that Doncaster does occasionally pay host to a circus, which pitches its tent close to the area of Charlotte's sighting.

  Personally, I can see that a large cat may have been deliberately released, or may have escaped, but it seems unlikely that the puma's mate would have "pined"; panthers are solitary animals, coming together purely to mate. Few mated pairs of animals "pine" at the loss of its mate. On the surface, it also seems unlikely that the circus would have taken the risk of releasing another large and expensive animal into the British countryside. That said, there is some precedent for this, and the story fits in with the all-round theory of just how large cats get into the wild in the first place. For example, a Yorkshire Post article (12 August 2000) describes how ex-lion tamer Leslie Maiden admitted releasing a panther and a cougar into the countryside off the A-57 Trans-Pennine road in the 1970s. It may be that this report became mixed up with an entirely bogus circus story, to become local legend.

- Diane also told me that the photographer from The Yorkshire Post mentioned that a year or so previously, at a farm "5 miles from Cusworth Hall", a "jaguar" leapt from a tree onto a farmer's combine harvester. The animal's claws missed the farmer, but scratched the harvester. This story was also told to me by a Doncaster policeman I discussed the case with afterwards, but he had heard that animal was a "tiger", and that it had scratched the farmer's arm.

I have a problem with this version, though. jaguars and leopards do climb trees, dragging their slaughtered prey high into the branches, away from scavengers, tigers rarely do so, and such is their huge size and weight. It is also my personal opinion that a tiger would have scratched the farmer's arm off, or at least made a horrific mess of it, but that's by the by.

- A worker from Cusworth Hall told Diane a local man kept big cats, including lynx, and a puma. According to the story, this individual had been jailed for drug dealing and that he released his animals into the wild before he was imprisoned.

This explanation is often heard, and now so common that it has attained elements of the urban legend, but there are definite precedents for this. Much as carrying guns is now part of the drug and gang culture, the desire to appear to their peers as ruthless, windswept and interesting in a Hollywood Villain sort of way has, according to RSPCA and police, driven many drug-dealers into buying exotic and more to the point, dangerous pets.

- Potterick Carr Nature Reserve, within walking distance of the Clarke family's home, has been the location of a couple of puma sightings in the past, and also the home of a small pack of wolf-hybrids, which during the Summer can often be heard howling in the night.

AFTERWARDS

Mark and I explored the fields and wooded area near the Clarke's home, which in itself is an expansive place, and borders Potterick Carr Nature Reserve, separated only by a fence and the train-lines.

Unfortunately, we found no physical evidence for big cats of any kind, or wolf-hybrids. The weather had been too dry and the ground consequently too hard for tracks. However, we found it easy to imagine that an animal like a large cat could exist there at least for a time. Finally, we returned to the field where Charlotte saw the jaguar, but - as we expected - the beast and any signs that it may have left were long gone. The tall grass had been cut, and the field ploughed over. However, the case continues: I have forged ties with the South Yorkshire Police, and have been added to their database. I have been assured that if any further big cat sightings are reported, I will be informed immediately.

The Knight, the Cat and the Poet; the tale of the Barnboro Wood Cat.
by Mark Martin

The Legend of Percival Cresacre's death has several variations, such as the events date. I have seen four different years proposed, however the date on Cresacre's tomb, in St Peters Church is 1477. When the legend began the name of the village was "Barnboro" and has evolved into "Barnburugh" over the centuries. There is an area, about three miles away, the supposed spot where the death fight commenced, known as "Cat Hill". Today the Village has a street "Cresacre Lane", named in honour of Percival.

The leading Fortean Andy Roberts, in his superb little book "Cat Flaps, Northern Mystery Cats" gives us his, fascinating, ideas on the mystery. This book is available from the CFZ on line shop and is essential reading, best read by torch-light whilst camping on a remote moor.

The earliest stories name the Knight slaying feline as a "wood cat" or a "Tom cat". According to Roberts (I'm sure he is correct) this was the name given to the wildcat (Felis silvestris), at the time, common in Yorkshire. There is a wooden effigy of poor Percy in St Peters, with a carving of the cat at his feet. The cat is about the size of a large domestic, or indeed, wildcat - much smaller than a puma or leopard. My personal interpretation is that a wildcat was involved, rather than a 15th century ABC. An extremely shy and not very large Felis silvestris would be very unlikely to attack a man and horse. If a man was attacked, I can't conceive how he would be killed. It would be no problem for him to see the cat off with the toe of his boot. My theory is that the wildcat scared Cresacre's horse and it threw him, causing his death. I think the epic battle story has grown up over the years, through exaggeration and to give the nobleman a more stylish death than falling from his mount and smacking his head.

On the tomb, in the Church, is an excerpt from a poem, "Esther's Tomcat". It was composed by a young, local man, who at the time was an up and coming poet. If interested in literature and poetry, you may recognise this fellow's name: Ted Hughes.

Sir Percival Cresacre effigy with the "cat" at his feet

THE GREAT CROCODILE HUNT
by Jonathan Downes

- CFZ YEARBOOK 2004 -

If my hero Hunter S Thompson - the father of gonzo journalism - could write a book called the Great Shark Hunt then I can damn well follow in his footsteps. Because I have never hunted for sharks, but earlier this summer I did embark on a great crocodile hunt - in suburban Staffordshire!

Charles Fort was an American philosopher who dedicated much of his life to cataloguing some of the world's strangest mysteries. After his death the study of such things became known as forteana. In one of his books – Lo (published in 1931) he discussed the mysterious appearance of crocodiles in the vicinity of a small town in the Cotswolds over a period of about thirty years in the mid-19th Century.

He wrote: "It seems to me that an existence that is capable of sending young butchers to medical schools, and young boilermakers to studios, would be capable of sending young crocodiles to Over Norton, Oxfordshire, England. When I think of what gets into the Houses of Congress, I expect to come across data of mysterious distributions of coconuts in Greenland!"

Sadly most scientists ignored Fort's findings and the matter was never satisfactorily resolved and indeed it has remained a perennial topic of interest amongst fortean zoologists ever since. In the early summer of 2003 it began to look as though history was repeating itself with accounts of crocodilians appearing at several locations across the West Midlands.

The CFZ (doing our best to live up to our claim that we are the best cryptozoological research organisation in the world) immediately dispatched two of our most experienced members to investigate. Andy Stephens went to the stretch of canal where the Gloucestershire croc had been spotted and produced a feasibility document which concluded that there could, indeed be a crocodile loose in the canal; while Mark Martin drove from Sheffield down to Cannock where he interviewed a number of witnesses and reported back to CFZ base much the same thing.

Now, we may be the best cryptozoological research organisation in the world but we are skint most of the time, and we had just sent my friend, colleague, and fellow contributor to this book, Richard Freeman, to Sumatra. Mounting an expedition to Indonesia is an expensive business and the CFZ coffers were pretty well dry. However, we had just about enough money to do a thorough investigation of one of the croc sites, so we flipped a coin and decided to go to Cannock.

We finally reached Cannock in the early afternoon of 21st July. After a rendezvous at our digs the Exeter contingent and Mark Martin drove in convoy to the pond at the end of Roman View. No matter how many times one carries out an expedition like this to finally see the location of a series of mystery animal reports for the first time.

The pond where the crocodile had been reported was surprisingly wild looking - an oasis of sanity in an increasingly desolate and unattractive West Midlands Environment.

- CFZ YEARBOOK 2004 -

John Fuller prepares the inflatable boat

On the far side of the pond from where we set up our temporary base camp, a new section of the M6 was under construction; furthermore, what looked as if it had once been virgin woodland on the hillside opposite had been flattened in order to build a featureless and rather nasty out-of-town shopping centre.

The ground immediately surrounding the pond looked rather more inviting. A wide range of butterflies and other flying insects fluttered, hovered, and buzzed their way around the thick vegetation, which was about 800 yards long and 300 yards across and was fringed by reeds and bullrushes. A contemplative looking heron sneered down at us from a large bush at one end of the pond, and - indeed – spent most of the weekend gazing down at us in a particularly supercilious manner. The pond was also home to a pair of swans and their three cygnets, who cruised up and down the water like majestic galleons and totally ignored us for the duration of our stay.

This was the largest gathering of CFZ personnel that had ever been gathered together for a single piece of fieldwork. From CFZ HQ in Exeter came me, Richard Freeman (who had only been back in the country for four days after his Sumatra trip), Graham Inglis, John Fuller, and Nigel Wright (on his first CFZ expedition for some years). We were joined by the aforementioned Mark Martin, Peter Channon (from the Exeter

strange phenomenon group), Chris Mullins (from Beastwatch UK), Neil Goodwin (from Mercury Newspapers), and Wilf Wharton (the CFZ Wiltshire representative soon to be emigrating to the Antipodes).

I split the available personnel into three field groups.

       1. THE BOAT TEAM (Mark and Graham)

       2. THE AWAY TEAM (Richard, Wilf, Chris, Neil and Peter)

       3. THE SHORE TEAM (Me, John and Nigel)

Even as John, Graham and Mark struggled to get our trusty dinghy onto the water, the first set of eyewitnesses arrived. They were a motley gaggle of teenage boys who came up to us, and in thick Brummie accents asked, "whether we were here for the crocodoile loike?"

The group of teenagers went about their business, and we went about ours. However, at least at first some of the other local residents were not as friendly. From the moment we arrived, the net curtains began to twitch and soon a procession of local residents walked past us - nonchalantly - to find out what we were doing. Nigel spent much of his time in conversation with these people, explaining the details of our mission and reassuring them that we were perfectly harmless.

Finally, we managed to get the boat onto the water and the away team were dispatched to the far side of the pond. At about 6.15, after a series of false alarms, Mark Martin in the boat had a sighting of what appeared to be the 18-inch long dark blackish green head of a large animal. It was not a positive sighting of a crocodile, but it was the best that we had managed to achieve. At the same time, the away team found an area of flattened reeds, which had looked as if a large animal had made itself comfortable after emerging from the waters of the pond. Unlike other such areas around the shores of the pond there were no downy feathers from one of the swans, and as the area of flattened vegetation was too big for any known mammal species from the area, it seemed quite possible that this had been the resting place for our mystery crocodilian.

Then in the early evening, John Mizzen, one of the original witnesses who had been interviewed by Mark Martin turned up. He told us, "It was about 5 ft long and that is including a tail of about two feet. Its head was flat, as were its jaw and its nose, and it was dark greenish black in colour and about 18 inches wide. The tail had a scaly appearance, and then it went underneath the water and we just lost contact with it. It had been on the surface for about three or four seconds and in that time it covered about 15 to 20 feet."

Another witness said that there been a series of incidents at a slaughterhouse which was on the shores of one of the other ponds connected to Roman View Pond by a water-

A warning poster designed by one of the local children

course. Apparently this establishment - which dealt predominantly with the despatching of elderly and ill horses - supplied meat to local zoos. Some of the meat was hung in concrete pits in order to prepare it for consumption by zoo animals. Whilst it was hanging something had taken enormous bites out of the carcasses.

Later in the evening, as it was approaching dusk, we returned to the pond and spent three hours searching the surface of the pond with three one-and-a-half million candle power spotlights. The away team, with head torches strapped on, scoured the bank, and out in the middle of the lake, Mark and Graham sat patiently in the boat waiting for a scaly monster to surface. Needless to say all these searches were fruitless and at about one in the morning we packed up for the night.

The next day was spent tying up loose ends. In the original newspaper report a local lady called Natalie Baker, was quoted as saying that her children and their friends had been so excited by the media activity following the initial crocodile sightings that they had spent some time making coloured posters of the animal as part of a school project. Now, Nigel has been working with and for me for nearly seven years now, and over the years I have asked him to do some extraordinary things for me. I have never before said to him "Dude, I want you to find me a little girl who draws pictures of crocodiles". But I did, and - not at all to my surprise, because over the years I have known him I have come to rely on his powers of deduction a great deal - he not only found me the little girl, but managed to persuade her to give me one of the aforesaid posters.

Flushed with success after that particular triumph, Nigel and I went off in order to try and solve another mystery, which - we felt - was likely to have a pivotal importance in solving the case of the Cannock crocodile once and for all. We had been told about the activities of a particularly unscrupulous reptile dealer who was - allegedly at least - operating in the Cannock area. Nigel and I left the shore party and the boat party doing their own respective things and went undercover.

It was surprisingly easy to track this fellow down. He had left a trail of debts a mile long, and wherever we went we couldn't find anybody who would say a good word about him. We found the shop where he had once operated a business, which - according to one of our informants - had been closed down on animal welfare grounds. We spoke to his erstwhile landlord and found that when he closed he had left large sums of money owing. We found that he had then set up business under another name in another part of town, but this too had gone the way of all flesh. After two failed businesses, we discovered that the person in question had most recently been sighted working part-time for a pizza delivery company, and selling the remnants of his stock through small ads in the local paper.

Although we cannot prove it, we were convinced that this discovery had essentially solved the provenance of the Cannock crocodile.

That evening we slowly began to break camp. John and Neil lit a barbecue, which had been donated to us by Chris Mullins and soon the fragrant smell of slowly charring burgers drifted over the evening wind. Someone produced the remains of a bottle of Scotch and Nigel appeared from Sainsbury's with two dozen bottles of beer. The CFZ drank, ate, and watched the sun go down. Neil disappeared back to Liverpool, and the rest of us went down the pub. Tomorrow was another day and we had another investigation to undertake.

Unfortunately we had not caught a crocodile. From the eyewitness descriptions, Richard and I are fairly convinced that we are talking about a spectacled caiman of between 3 and 5 ft in length. Sadly - unless it is very lucky, and somebody manages to fish it out of one of the connecting streams - it is doomed to a slow and ignominious death as soon as the first chills herald the advent of the season of mists and mellow fruitfulness. And all because of some stupid selfish idiot who wanted an exotic pet!

C'est la vie.
Unfortunately.

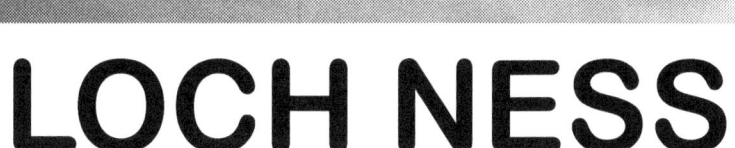

LOCH NESS
Steve Feltham Interview
by Mark Martin

- CFZ YEARBOOK 2004 -

Mark Martin, Lincolnshire representative for the Centre for Fortean Zoology, and long-term member of the Permanent Directorate has been interested in the Loch Ness Monster phenomenon for many years. Here, he travels to the great lake, and interviews one of the most dogged Nessiphiles of the current generation: Steve Feltham.

MARK MARTIN: "So, Steve can you tell us how long you've been in Loch Ness?"

STEVE FELTHAM: "Eleven years, just over; June 1991 I got here."

MARK MARTIN: "And you mentioned, a while ago, you had a sighting up near the other end, something like a torpedo?"

STEVE FELTHAM "Aye, it was May 1992, I would guess about this time of year, it was just going against the waves as if a torpedo was going through there, all you could see was a streak of water, a white line going against the waves, a splash of water as it hit each wave you couldn't really tell what it was at the distance of the waves so - no explanation for it as far as I can tell. There is no fish in here that could have caused a disturbance like that. I don't know what it was."

MARK MARTIN: "You said it was at the other end?"

STEVE FELTHAM "Aye, that was at Fort Augustus."

MARK MARTIN: "So out of all the evidence available, photographs, sonar tracings and eye-witnesses reports what do you think is the particular strongest?"

STEVE FELTHAM: "The eye-witness reports are the backbone of the whole belief here, and it's not the visitors in the area that see things that keep me convinced. It's the people that live in these villages."

MARK MARTIN: "I've noticed that. I can tell. I've been up there."

STEVE FELTHAM : "I've been up here for a decade or more you know and they've been here for 30 years , they're used to seeing all the wind action, false alarms, the things that fool tourists so it's that which makes eye-witnesses that come up here all the time they come and tell me because they know why I am doing this."

MARK MARTIN: "You've been here 11 years, how many reports have you had from local people?"

STEVE FELTHAM: "Numerous, probably a couple of dozen good ones. I also get a lot of tourists turning up with bits of film and stuff which, if they come through me I can manage to back them off and tell them no that's not what we're looking for but if they go through other outlets then their stories sometimes go round the world with the latest or greatest photographs or whatever which we all look at and say that's a boat wave or a stick. I try and keep those, if they come through me."

MARK MARTIN: "And seeing, obviously that you are convinced there are animals in there, the question that a lot people ask, a lot of scientists ask about and are curious about, is what particular type of animal is it? Is it a pleisosaur, or some form of mammal, or some form of giant eel?"

STEVE FELTHAM: "It's a mystery. I don't know what they are, sorry. You can dismiss a few things, you can dismiss plesiosaurs. I am convinced there are not plesiosaurs in there because Loch Ness was only formed a long time after they died out. Well, there's that, and there were so many ice ages various over the last 70 million years, it's so unlikely to be plesiosaurs."

MARK MARTIN: "There's got to be something but it can't be something that's air-breathing."

STEVE FELTHAM : "Well, you can't automatically say that because we do get seals coming here, and certainly if they come past here I'll spot them. If they go near a fish farm the guys there will spot them and they're in here, maybe not every year, but maybe say you get one a year. You'd think tourists would come along and say right that's the mystery solved, I've seen it and we'd say it's a blooming seal, but tourists don't spot these things, they're air breathing and their curious animals that are on near shore. So if they can't spot a seal then maybe they can't spot an air-breathing whatever if only a small portion of the head would stick out of the water."

MARK MARTIN: "Thanks very much."

IN SEARCH OF GIANT BATS
by Richard Freeman

The Order Chiroptera, or 'Bats', is broken into two distinct groups. Microchiroptera - the micro-bats - are generally small. They feed mainly on insects but some (depending on the species eat fish, small vertebrates, nectar, fruit, and blood. The biggest of them is the Australian false vampire bat or ghost bat (Macroderma gigas) with an 80cm wingspan.

Megachirdoptera or fruit bats feed exclusively on plant matter and are far bigger. They are easily recognisable by their dog like faces. The biggest is the Malayan flying fox (Pteropus edulis) and has a wing span of 1.5 meters.

However, there are reports of what sound like gigantic micro-bats that dwarf even the largest flying fox. Here, I will take in a whistle-stop tour of the tropics and relate tales of these monstrous bats that still allude science.

It is fitting that we begin in Africa, the Dark Continent. Still untamed in the 21st Century, there are areas of jungle larger than Scotland that have never been properly explored. From these dark, primal recesses come tales of gigantic bats.

The late Ivan T Sanderson was a cryptozoologist, author, and explorer. Whilst in the Cameroons on an expedition led by Percy Sladen in 1932-3 he encountered a giant bat in the Assumbo mountains. He was hunting along a steep banked river one night and had just shot a fruit bat. The creature fell into the river and Sanderson attempted to retrieve it. He lost his footing and fell into the water. Then one of his companions shouted at him to look out. Sanderson takes up the story.

"And I looked. Then I let out a shout also and instantly bobbed down under water, because, coming straight at me only a few feet above the water was a black thing the size of an eagle. I only had a glimpse of its face, yet that was quite sufficient, for its lower jaw hung open and bore a semicircle of pointed white teeth set about their own width apart from each other.

When I emerged it was gone. George was facing the other way blazing off his second barrel. I arrived dripping on my rock and we looked at each other…

…Just before it became too dark to see, it came again, hurtling back down the river, its teeth chattering, the air "shss-shssing" as it was cleft by the great, black, Dracula-like wings. We were both off guard ,my gun was unloaded, and the brute made straight for George. He ducked. The animal soared over him and was at once swallowed up by the night."

Natives later told the men that what they had seen was an "Olitiau" a type of giant bat.

Gigantic bats were once reported in what is now Ghana and Togo in West Africa. Sasabonsam is a demonic creature much feared by local tribesmen. It is said to be covered in long red hair and be aggressive to humans. They believe that it is friends with sorcerers and witches.

A. W. Cardinall, in his book Tales Told from Togoland, recounts how he was told of the killing of one of these giant bats by the son of the hunter who slew it. A man called Agya Wuo killed one of the monsters and brought it into town. A scaffold was built to display the horror. He recounted…

"It is taller than myself, very, very tall. The ankles and feet are very, very narrow. The hands are long and can be stretched. When it was first brought it was not quite dead, and it made noises. "Ho ho". When it died the hands got pulled in, and only a small portion (the palm and fingers) could be seen.

It has fingers like a human being. The forehead is very smooth. There is an abundance of hair on the head like a woman's hair, but hard or stiff. The claws are very hard and very long. It has wings like a bat, and they can be stretched out into the "lorry" road. (over 20 feet)

The wings are very thin, but it is not easy to tear them, unless by cutting. There is plenty of hair at the back, but on the stomach not so much hair. There is plenty of hair at the sides from the armpits down. The legs were straightened out when I saw them, but it was said Sasabonsam could twist them around a tree. The arms are very long, reaching downwards to the feet.....

…The teeth are very long, like those of a dog, sticking up and down. The skin is spotted black and white. It has no heels, but the leg continues to the sole of the feet. The hands are the same, the palm indistinguishable from the rest of the arm by any noticeable break. It has short horns. Two of them. The lips are like those of a human being, but wider, more stretched at the sides. You can see big bones at the sides, very big bones.

The chest is not like that of a human being, but like that of a fowl, with a ridge at the centre. The Sasabonsam was rolled into a bundle and tied together to be conveyed to Kumasi. The man said the animal was sleeping in the hollow of a tree and crying. The cry was like that of a bat but deeper. It had a long beard. The nose was like that of a man but more prominent. The man hearing that the animal was crying came home to prepare for shooting it. He had seen part of the animal, a hand hanging from a hollow at the top. He could see however, it was not a human being's hand.

Coming home, he asked advice and loaded his gun with certain things. He did not use ordinary pellets. Mr L Wood the District Commissioner, came there and the body was taken to a bungalow and photographed . It was 1928. February 22nd.

Not all of Africa's mystery bats are massive. A monster can be a monster even without huge size, as recounted by archaeologist Byron de Porok in his 1943 book, Dead Men Do Tell Tales. Shortly before Ethiopia was invaded by the Italians in World War 2, Porok was on an expedition to the country's southern region. He was intrigued by the stories he heard of the much feared "Devil's Cave" near Lekempti in the province of Wagala. It was said to be haunted by horrific winged entities called death birds. Most of the natives were too fearful to show Porok the way, but after a substantial bribe one man agreed to guide him there.

The cave was located high in forested hills. Porok gallantly braved its dark recesses and discovered that the death birds were, in fact, small bats. On returning to the surface he asked two local goat herders why the small bats were called death birds and why they were so feared.

The men explained that the bats fed on human blood that they took from sleeping victims at night. When they saw the white man's scepticism, they showed him wounds and bite marks on their arms where the death birds had bitten them and lapped up the blood. Eventually the nightly feeding took its toll and the victim weakened and died and had to be replaced by new herders. They also took him back to their camp to show him a victim.

Close to death from the bat's nocturnal feast, he was little more than a skeleton wreathed in ashen skin, too weak even to stand. He lay in a child's cot, his clothes in a pile beside him were blood-soaked rags.

All three species of known vampire-bats hail from the new world. They inhabit south and Central America. No known species of old world bat feed on blood. If the stories were true then the Ethiopian death birds would be a brand new species of bat. As yet no zoologist has returned to Devil's Cave to find out!

There is another explanation. A few years ago a terrible outbreak of the Ebola virus was traced to an ill-explored cave in Uganda. The virus seemed to have had its genesis or at least its gestation deep in the cave. Ebola causes blood vessels to break down and the victim to bleed from their pores (much like Poe's fictional Red Death). Could the blood loss around Devil's Cave have been an outbreak of Ebola or something akin to it? Perhaps the bats were just scapegoats taking the blame for a new disease the locals could not understand.

Tropical Asia also has its own legends of gigantic bats. On western Java the Sundaese people tell of seeing a gigantic type of bat they call an Ahool. The name refers to the creature's ghostly cry. The Ahools are said to have 12 foot wingspans, bodies as large as a one year old child, and a covering of grey fir. The face resembles that of a monkey. The Ahool's feet appear to be turned backwards. This strange feature makes zoological sense as they are believed to feed mainly of fish. Other fish-eating bats such as the bulldog bat (Noctilio leporins) have feet shaped like this to facilitate scooping up their slippery prey. They roost in caves behind waterfalls in the day and emerge to hunt over water at night.

The Ahool is said to be wary of humans and tries to avoid them. But there is one story of a native who was attacked and wounded on the arms by one. Dr Ernest Bartels, a European educated resident of Java has collected and preserved accounts of the creature. He himself had heard its distinctive cry "Ahoool, ahoool, ahoool!" As with Sanderson's Olitiau, the flat monkey like face of the Ahool suggests a microbat rather than a fruit bat with its dog-like snout.

An even larger creature is reported from the island of Saram in the Moluccas. In June 1986 tropical agriculturalist, Tyson Hughes visited the island and collected detailed reports of a giant bat like creature said to inhabit its forests. The Orang-bati are human like in form 4-5 feet tall, have red skin, black wings, and a long thin tail. They emit mournful wails and the native folk go in fear of them. The literal translation means flying man. They are rumoured to roost in extinct volcanoes and at night to fly out across the forest to the coastal villages and abduct children.

Is there any truth to the legend of a giant child-eating bat in Saram? Perhaps we will soon find out, as, despite being declared a nature reserve, Japanese logging companies are fast stripping Saram of its rain forest. If they exist the Orang bati may soon have nowhere to live.

To wrap up our exploration of giant bat stories, we move out of the tropics to that vast frozen forest known as the Russian tiga. Siberia is one of the least inhabited and poorly explored places on earth. It abounds with strange tales of monsters. Lake and forest dwelling dragons, ape men, giant prehistoric bears, even surviving mammoths! Yet stranger still than all these is the Letayuschiy chelovek like Orang bati this translates as flying man. They are reported from the Primrskiy Kray Territory in the Russian far east. Several years ago hunter, A. I. Kurrentsov, reported one swooping low over his campfire. He shot at it as it flew into the night. He described its cry as like a woman's scream but ending in a lugubrious howl.

Do these monsters inhabit the tangled forests and deep caves of the unexplored regions of our world or are they products of the equally mysterious meshes of our own minds? Only dedicated expeditions will find out. With the current apathy of scientists, and film-makers, the riddle of the giant bats will go unanswered.

CHUPACABRA
Jaime Maussan Interview
by Jonathan Downes

In 1998 Jon Downes and Graham Inglis from the CFZ travelled to Mexico in search of the grotesque, vampiric, chupacabra. While there, they interviewed Jaime Maussan, the doyen of Mexican UFOlogy, chupacabra, and fortean research..

JON DOWNES: Can you tell us a little about the history of the Chupacabra attacks here in Mexico?

JAIME MAUSSAN: They started on July 17th 1994. We have papers that show that then people were going out into the mountains to search for an animal that was sucking the blood from other animals - especially cattle. At that time they didn't know the name chupacabra, which is important because it shows that this is a real phenomenon and not something invented by the media.

JON DOWNES: What sort of animal did these first witnesses describe?

JAIME MAUSSAN: They were describing an animal that they said was probably feline. But in the very first report that they said that it could also fly. It wasn't until 1996 when we heard the name chupacabra and since then we have had hundreds and hundreds of reports.

JON DOWNES: In Mexico you have both pumas and jaguars. Were these sightings of a 'feline animal' in places where your indigenous big cats naturally live?

JAIME MAUSSAN: Not in recent times. The places where these sightings took place were where there have not been any species of big cat since at least the beginning of the century. Some people still talk about them but there is no evidence that they are still there. First of all, I am an investigative journalist, and I have been doing my investigations into UFOs since 1991 because we have had many sightings since then, and I have been reporting them consistently ever since and have amassed a large body of data. When these incidents started to happen I became involved in investigating them with the same objectivity as I did the UFOs. I soon found that this was a real phenomenon and one that people were afraid of even more than they had been of the UFOs.

JON DOWNES: How much personal involvement did you have in the research?

JAIME MAUSSAN: We have been to the places where these attacks have occurred and we have spoken to the people there. We have had many reports from the different states - and we have built up a network of co-operating investigators and we have been able to establish what has been going on all over the country over the last couple of years...

JON DOWNES: What sort of animals have been attacked?

JAIME MAUSSAN: Mostly sheep and sometimes chickens and rabbits. Far more seldom, bigger animals like cattle or donkeys. At least 80% of the time I would say that

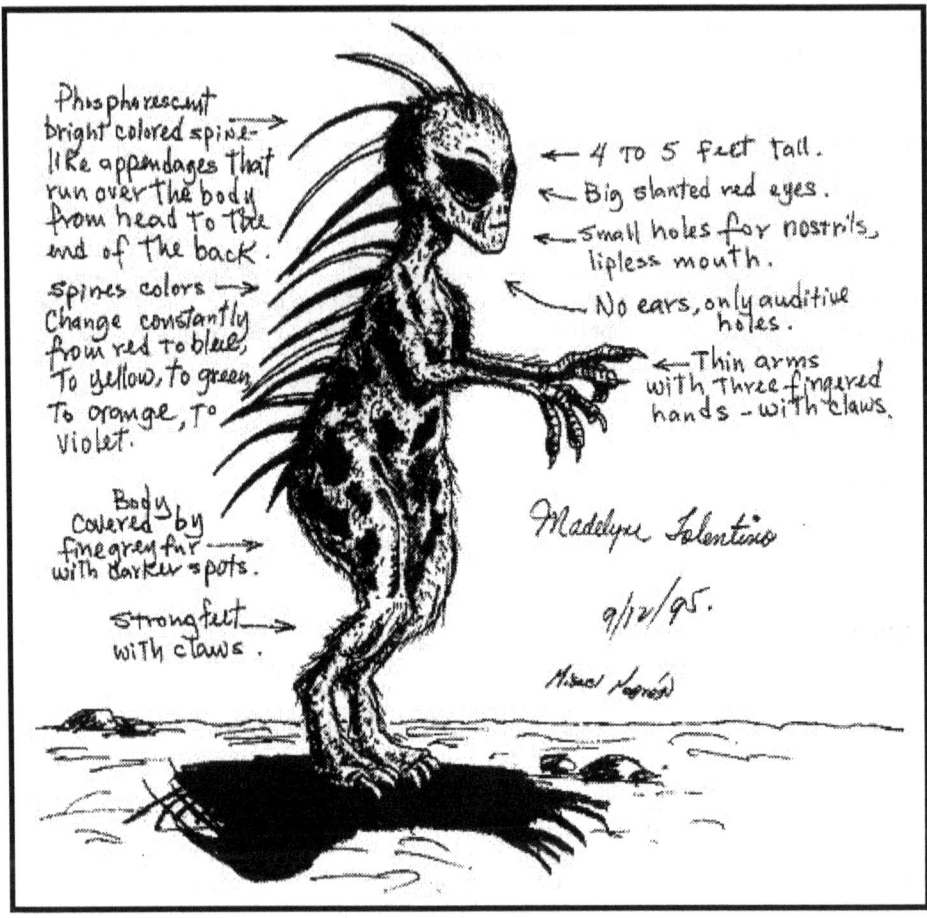

This drawing of the notorious Goatsucker, also known as el Chupacabra, was provided by eyewitness Madelyne Tolentino in 1995, and includes detail descriptions of various purported physical features of the mysterious creature.

the victims have been sheep.

JON DOWNES: Are they always domestic animals rather than wild ones?

JAIME MAUSSAN: Yes. It is always animals related to man and that is easier to understand because they are closer and easier to report than wild ones. They are also larger. I think that one of the reasons that the chupacabra attacks near man, is because it is easier to find their food.

JON DOWNES: Have there been any reports of deer or other large wild animals being attacked?

JAIME MAUSSAN: Not in Mexico, although we have reports of deer being attacked in San Antonio in the United States. On this occasion there were ten deer attacked. Even then, they were wild animals in a controlled situation and a number of them were attacked. It seems that wild animals are somehow more protected from this kind of presence. They can escape more easily and can probably hide from the predator more easily.

All I can tell you is that this phenomenon is real. We have several cases where animals have survived for several hours after having been drained of blood. We also know of several cases where humans have been attacked by these beings. We also know of several very close encounters and things that tell us that this is not natural. I cannot tell you exactly what it is, but I can tell you that it is true.

JON DOWNES: Is there a history of stories about Vampires in Mexican folklore?

JAIME MAUSSAN: Yes. We have many stories of that kind that are very well known across Mexico. They are not really from this century, but from the past. They tell of people who were once humans who were turned into these `animals`, which were able to suck the blood from their victims. They are similar to the ones that you have in Europe, but there are many little differences that tell you that they do not have the same origin.

JON DOWNES: Why is that?

JAIME MAUSSAN: I don`t know. It would be interesting to investigate them further because it suggests that they have their origin in fact rather than in memories of stories brought from Europe. It may mean that something like the chupacabras had happened years ago. Remember that in Mexico the phenomenon is different from that in Puerto Rico because we have had attacks on humans. These are very well documented with solid evidence. In one case, the person who was attacked had to fight the creature in order to escape and he was in a state of shock. The events were witnessed by his wife and brothers. All of them tell exactly the same story. The creature smelled very bad, and the victim was in a state of shock and had two holes with blood in his arm. This is a typical story when you examine the accounts of such figures from previous centuries in regards to blood sucking attacks. Perhaps this is the real origin of these stories. Perhaps these vampires were NOT humans, and it is the same old creature that has been with us for a long time and we have never been able to discover exactly what it is.
It seems that this `creature` is migratory. It stays for a while in a district and then leaves - perhaps coming back again years later.

JON DOWNES: Some people have linked these attacks with UFO reports.....

JAIME MAUSSAN: You have elements to suggest that these `creatures` are coming from another world. It could be. There are also suggestions that it comes from another reality or dimension but we do not have enough elements to say that it IS. I can tell you that after having investigated this phenomenon for more than two years that it is real, it is a fearsome phenomenon, and having been in places where there has been a spate of attacks I can confirm that it is like no other animal.

JON DOWNES: How do the communities react after there have been a spate of the attacks?

JAIME MAUSSAN: In one town, they painted crosses all over the walls of the buildings, especially in the direct vicinity of the attack. They thought that if they did that they would somehow be protected. In Sinaloa, the whole community was so frightened that the army had to be called in to give protection because the people couldn't sleep any longer. They couldn't go out because they were frightened. There were so many attacks on people that the community was afraid for its children. They figured that if animals could be attacked then their kids would be next.  Animals were disappearing each night and people were afraid that soon it could be a little boy.

The attacks were always at night. We do not have any reports from Mexico of attacks having happened during the daytime...

JON DOWNES: Have any scientific studies been done on the corpses left by the `creature`?

JAIME MAUSSAN: We know that there is something in the holes, which seems to allow the blood to flow rather than coagulating. We have also found some other strange elements about the corpses. However, the sort of work that really NEEDS to be done has not yet been carried out in Mexico. Some of the Universities are interested in carrying out studies. Some physicians are also interested, but unfortunately I would say that so far there has been very little scientific work done.

JON DOWNES: You have told us how the Government, the police and the army have reacted to these attacks. How has the established Church reacted?

JAIME MAUSSAN: Not a word. I think that in the older times they would have blamed these attacks on The Devil, but now they do not want to say that The Devil is here! They have given no official answer.

JON DOWNES: Have priests tries to carry out exorcisms?
JAIME MAUSSAN: No. This is because everyone believes that it is NOT the Devil

but some kind of an animal.

JON DOWNES: Do all the witnesses describe the same type of animal?

JAIME MAUSSAN: The recent sightings are very consistent. They describe an animal about one metre tall with a face like a mouse or a kangaroo with little hands, very thick feet, wings, a small spiny back, and some times it is described as flying...

JON DOWNES: You said that the early attacks were of some sort of big cat with wings. Are these creatures still being reported now?

JAIME MAUSSAN: Those were the first descriptions that were received before the media started to present reports about chupacabras. That happened in Puerto Rico and has now spread to Mexico. However, all the people who have described the creature have seen it very closely - some times from less than a metre away.

JON DOWNES: Is there any physical evidence of the creature? Hair samples for example?

JAIME MAUSSAN: No. There are some traces of what appear to be footprints and there have been samples found of what appear to be the excrement of the animal. However we are ready now. If these attacks start to happen again then we will be ready for them. In Mexico there are enough professionals interested, that we should soon have the answers that we are looking for.

The problem, both in Mexico and in other places in the world where similar events have taken place, is that these incidents have taken place far out in the country where there are no facilities to carry out an investigation. When you find out about the incident it is usually two, three or even four days after it occurred and by then it is too late. You have to be there immediately.

JON DOWNES: what do you intend to do next?

JAIME MAUSSAN: When the Mexican incidents started in 1996 they took us by surprise. They started to happen in the January and by the middle of April they had attracted the attention of the media, and we became aware of the problem, and by the end of May the main spate of attacks were over. By the time we had everything set up in order to investigate the attacks they were over.

The most important series of attacks were in Sinaloa. There you can find many people who have been attacked, and the collective fear of the community is remarkable. It was there that we had most reports of attacks on humans.

IN THE WAKE OF THE MONKEY MAN

by Jonathan Downes

- CFZ YEARBOOK 2004 -

In the opening months of 2001, just as the United Kingdom fell under the collective horror of the foot-and-mouth disease epidemic which scarified the rural communities and left when the entire country under way (an often literal) pall of smoke from the massive funeral pyres, strange events were taking place halfway across the globe in India and Pakistan, where entire communities were under siege from a monster that became called The Monkey Man. This "creature" - for whatever it was, it is almost certain that it is not an animal at least in the way that we understand it - became so ubiquitous that it even featured as a throwaway line by Apu's fiancee in an episode of The Simpsons.

Some described it as being a bigfoot-like creature, some as having a metal claw or claws, while others likened it to a cat with glowing eyes. Another claimed it had flaming red eyes and that green lights glowed upon its chest. Other witnesses described it as agile and feline, as a bandaged figure, or as a helmeted thing.

One of the most disappointing things about being editor of the world's only cryptozoological magazine is that - being limited to a 60pp format, we're often not able to give the space to subjects that we would like to. Therefore, although we mentioned the Monkey Man episode in the news pages of Animals & Men, it was in passing, and we were unable to give it the space that it deserved. Following the positive response that we have received to a paper I wrote in the 2003 Yearbook, in which I presented our collection of source material about the 2001 cattle mutilation episode in South America, together with my notes and comments, we have decided to do the same thing with our monkey man archives in this present volume.

Interestingly, although this first news item - from The Hindu of 17/08/2001 describes the panic as having been going on for over a month, this is the earliest news item that we have found on the subject:

COLLECTIVE FEAR - MONKEY BUSINESS.

by Lionel Tiger.

The extreme often illuminates the normal. So the bizarre reports of the "Monkey Man" which has, beginning April 8, terrorized people in the Indian province of Utter Pradesh and then in New Delhi, may reveal what lies beneath the surface of everyday life.

Self-satisfied sophisticates may fend off the case by placing it in the category of primitive magical thinking. However, it is worth exploring what this eruption of hysterical fear meant for the people who experienced it, as well as for our own community - one in which the Loch Ness Monster, Big Foot, New Mexican space visitors, and the Yeti compete for tabloid sales with details of Cher's latest makeover.

What happened in India was that flash announcements swept the community that a half-man, half-monkey creature was randomly attacking people at night in the streets where many of them slept because of the heat.

After complaints about 348 attacks by the monster, vigilante groups wielding sticks patrolled the streets. Several hapless candidate-monsters were attacked. Three thousand police officers were assigned to find the creature. At least three people are thought to have died by jumping from windows when the monster pursued them - or perhaps their enemies took the opportunity to push them out. A police reward of 50,000 rupees (around $1,000) turned up no useful leads. It was finally announced that the monkey-man was the product of "fear psychosis." In any case, we can be sure that he will not be captured because he does not exist.

But that is not the central point. The point is that monkey-man taps into a deep vein of interest in the ghoulish and disastrous, not only among Indians but also Westerners. Highway police know that after an accident people will slow down to gaze with evident fascination at someone else's wreckage. One of the most common television news images is the yellow police tape marking a crime scene. Morbid curiosity is surely a prime ingredient of the appeal of dozens of films, television series like "The X-Files," and mystery thrillers, to say nothing of those wholly freakish films and TV series featuring human actors who wear fearful makeup and costumes.

In short, consumers often seek out drastically unpleasant stimuli and pay for the privilege to boot, with time if not money. The monster monkey-man isn't very far away both here and in New Delhi. Even when it appears to recede too far, we contrive to restore its power through nutty fear psychosis on the one hand and well-crafted works of communication on the other.

What appears to be going on is that, however comfortable and predictable life may seem to people, the awareness remains that everyone from New York to New Delhi lives on the edge. Drive down the highway and mistakenly turn the steering wheel 10 degrees to the right or left and you may be a bloody mess or dead. The day-trader gesticulating to his cell phone on the turnpike has the speed and poundage to kill you in seconds. Children are taught to avoid walking under ladders with good reason - tools fall.

There are inner monsters, too, which connect the smooth modern present with the unruly dangerous past before excellent medicine and lawsuits for remedial damages. Nightmares don't only occur during sleep.

Before amniocentesis became commonly available, bearing a heartbreakingly deformed infant was neither a theoretical fear nor a rare event. There have always been tacit and sub rosa implications that somehow understanding doctors would fail to assure the survival of severely irregular newborns. Pregnant women were, and are, understandably frightened of everything from the Evil Eye to possibly toxic foods. Some element of the public turmoil about cloning and genetically modified foods must have to do with long-standing fears about genetically modified humans - monkey-men of our own devising.

The police say the creature is 4'6'', wears only a dark coat of hair.

Small children threatened with having to sleep by themselves seek endless reassurance and comfort. They lack certainty that the boogie-monkey-man will remain in its cage - perhaps this is why Western society is aberrantly alone among the communities of the world in requiring little children to sleep alone, in the dark.

And it remains a favourite tactic of degraded statecraft for a group in power to define a subordinate one as a kind of non-human monkey-man, perhaps by requiring it to wear a version of species marker such as the yellow star for the Nazis. Now we have the Muslim Taliban's up-to-the minute innovation, requiring Hindus to wear specially coloured garb, like clowns. They are saying: "We fear you because you are different. But we will prevail and box you in. We will make a monkey-man out of you yet."

Apart from the fact that the name of the author of the piece is either an extraordinarily funny lexilink, or a rather fatuous pseudonym, it is the tone of the piece which is most interesting. The author seems more intent on stressing that he is not superstitious enough to actually believe in such monsters, than he is in describing the reports that he has received. It is interesting that even in this - the first of many newspaper reports in our collection - the author is propounding the socio-political explanation for an outbreak of Zooform activity, that I have done, not only in the aforementioned article about the 2001 cattle mutilations in South America, but in my books The Rising of the Moon and Only Fools and Goatsuckers.

However, he is so intent on stressing that his country is not a nation of superstitious peasants that he almost protesteth too much.

During May, dozens of individuals were said to have been hurt in its attacks, and two took fatal leaps because they heard the "monster" was nearby. In daily newspapers, photos of scratched victims upset New Delhi residents.

Early during the flap, on the night of May 14th, fifty attacks were reported, according to the May 17th edition of The Australian. (Numbers vary widely in news

Eyewitness says it is 5'6'', wears black and sports a helmet with shining red eyes

19 May 2001 INDIA:

A Hindu priest chants religious slogans near a photograph of the Hindu monkey god Hanuman (top) during special rituals performed to rid the community of a mysterious ape-like creature in New Delhi on May 19th, 2001. The Indian capital has been in the grips of a mass hysteria over a mysterious ape-like creature that has allegedly clawed and bitten dozens of people, striking terror in the capital. Police said on Saturday they expected a breakthrough in the case soon.

reports.) The scare had at this point moved from its origins in Ghaziabad to a number of areas in East Delhi.

Early May 15th, at 2:30 a.m., a pregnant woman in East Delhi fell down some stairs after being awakened by the shouts of neighbours saying that the monkey had arrived. She died in a hospital, having been one of the two aforementioned jumping fatalities. That same night (late Monday/early Tuesday), police received 13 distress calls from the New Usmanpur area.

As of May 17th, police in Delhi had taken more than 40 - perhaps as many as 65 according to other accounts - calls reporting depredations of this alleged Monkey Man since May 13th, from many sections of the city.

However, the next clipping in our collection - from the 25th May - still shows the authors determined to see this as a curious sociological episode, which - bemusingly to them - had attracted an enormous amount of unwanted interest from newspapers around the world:

MONKEY MAN KEEPS LONDON TABLOIDS GOING.

by Rashmeez Ahmed and The Times of India News Service.

LONDON: After 10 days of mounting excitement here over Delhi's monkey man, there is consternation at his sudden and unheralded disappearance. "Is it over? Is he dead? Has he been captured? Or is he simply having a bit of a rest before resuming the reign of terror?" wondered Britain's most popular tabloid, The Sun, in the latest of a succession of close-up accounts of the "fiend who made a monkey of cops".

The paper also continued its earnest appeal to all readers to e-mail any sightings of the monkey man, even as it sadly pondered if the metal-clawed creature ever existed at all.

The Sun's pithy accounts were supplemented earlier this week by Channel 4, with the respected mainstream television channel's plans for a breakfast-time discussion on what one producer described as an "extraordinary story". She said the newsroom had been following the Monkey Man's growing catalogue of exploits through the virtual newscaster, Ananova. "We watched it grow and grow and we thought it was time to look at it, to find out whether it was just mass hysteria or something like the Beast of Bodmin".

The aforementioned beast, of course, was part of the pleasurably scary lore built up in parts of England's West Country. It featured great cats, some as tall as six feet, wandering around even as some farmers reported sheep being found messily killed.

For years, the Beast of Bodmin assumed an almost mythical place in a country that does good business out of deep mysteries such as the Loch Ness Monster, which is a hugely popular tourist draw in the Scottish Highlands.

The Channel 4 attempt at understanding the phenomenon involved a discussion over the cornflakes with the Fortean Times, a paper which regularly reports strange and wondrous news. It recently conducted an opinion poll that found one in three people in Britain believe Jesus literally rose from the dead. But the Fortean Times was less credulous about the monkey man, Channel 4 told The Times of India, adding that the paper counselled caution because there seemed very little "investigative material on the monkey man and certainly not enough to make a definite decision".

But the British tabloids seemed less considered in their response to Delhi's mass hysteria, offering helpful blood-chilling nuggets of information on a "similar panic in Latin America in the late 1990s, when there were reports of a mysterious, vampire-like beast that sucked the blood of livestock and was named the Chupacabra or goatsucker".
The broadsheet newspapers, however, remained more restrained, even as The Times

continued the theme in a minor key by marrying India's beastly tales with the infuriating disappearance from public life of its other export, embattled government minister Keith Vaz. It carried a cartoon that showed two people chatting even as the Abominable Snowman hovered in the distance. "Look there's the Yeti," said one, "Ask him if he's seen Keith Vaz".

It was both amusing - and for me at least, culturally interesting - to see the authors of the above piece critiquing the British media. Again, the tone of the article sees the authors desperate to pooh-pooh any suggestion that India is a primitive nation that needs to be dragged - kicking and screaming - into the 21st Century. In a stark contrast to the South American newspaper reports about the cattle mutilations, which included some of the most bizarre hypothesising that I have ever read, linking the events to putative UFO activity, and Government Cover-Ups, these reports are almost clinically sober.

However, in a complete opposite of the time frame of events in South America, the clergy got involved publicly before the police:

Source: TIMES OF INDIA 20/05/2001

19 May 2001 INDIA: HINDUS MAKE RELIGIOUS OFFERINGS IN NEW DELHI TO RID THE CAPITAL OF A MYSTERIOUS APE-LIKE CREATURE.

Hindus make religious offerings into a holy pyre during special rituals performed to rid the community of a mysterious ape-like creature in New Delhi on May 19, 2001. The Indian capital has been in the grips of a mass hysteria over the "ape man" that has allegedly clawed and bitten dozens of people, striking terror in the capital. Police said on Saturday they expected a breakthrough in the case soon.

Four days later - on the 23rd of May - the Times of India printed the first official response from the police:

MONKEYMAN NOT AN ANIMAL - INDIAN POLICE.

New Delhi police think that the ape-like creature, which terrorised the city over the past week, is a human. Dozens of people in the Indian capital have reported injuries caused by the mysterious creature dubbed the "monkeyman".

Three frantic people have fallen to their deaths from buildings so far because they thought monkeyman was chasing them. Police have yet to solve the case despite narrowing down suspects in the week-long saga.

Police said they had received 324 complaints of people sighting or being bitten and clawed by the creature, 260 of which were found to be hoaxes. They added that medical reports showed that none of the injuries of the 64 people hurt in the incidents were caused by nail or tooth marks. Delhi police have offered a reward of 50,000 rupees (US $1,063) for information leading to the capture of the "monkeyman" which they now believe is not an animal but human.

Descriptions of the attacker have varied wildly. Some people have said it had a metallic claw, others that it was like a cat with glowing eyes and one said it had flaming red eyes and green lights glowing on its chest.

On the same day the first of the major international news agencies joined in with this report from MediaCorp News.

ON THE TRAIL OF AN UNCANNY CREATURE.

by Kenneth Wright.

CONNOISSEURS of what American newsmen call a Giant Pumpkin story - the idea being that when hard news is thin on the ground, a prize-winning prodigy of squash or succotash in South Carolina can get a show on an inside page of the Washington Post - will have shared my keen interest in the Monkey Man of New Delhi, an uncanny creature who has been manifesting himself and disappearing in mysterious ways, all to what Scots law - if my juvenile citation for breach of the peace is to be believed - calls "the fear and alarm of the lieges". The Monkey Man is reportedly anything from 4ft to 7ft tall, has red glowing eyes, and is covered in thick fur or naked as a jaybird, according to who saw him last. The question of whether or not he has wings and the power of flight is equally controversial among his many witnesses.

Although he has yet to do anyone any corporal harm, New Delhi is practically under curfew and the police are under orders to take the Monkey Man dead or alive, with no great official preference in the matter. His capture I await hourly, but I'm not holding my breath. What we have here is obviously a case of mass hysteria, but I always wonder on these occasions whether the hysteria is really as mass as all that. Suggestibility is a powerful thing, but, then, so is that aspect of human nature summed up by my hero Brendan Behan (Ireland's greatest drunk writer and greatest writing drunk) in the words "You might as well be out of the world as out of the fashion". No-one wants to be the only chap on the street who hasn't seen what all his neighbours have.

Most UK reports of the New Delhi story acknowledge this in a between-the-lines sort of way, which is fair enough by me; but they also - even in the ultra - correct Guardian - have a touch of the Funny Foreigners about them, as though seeing imaginary monsters were the prerogative of Rudyard Kipling's "lesser breeds without the Law".

Not so, not so. Well within living memory and in my own city I can cite some very similar phenomena. Like every slum- tenement Glasgow child who had the ill-luck to live on the ground floor, where the communal lavatory was at the back of the close, en route to the back court, I lived in nightly terror of Flannel Feet, a vague but terrible and soundly verified entity who abducted and ate small boys answering the late-night call of nature. And then there was that dreadful day in the mid-1950s when hundreds of Glasgow children, armed with improvised weapons, invaded the Necropolis cemetery in search of a creature variously reported to be Dracula, Frankenstein's monster, a robot, or an alien spaceman with metal teeth (and red, glowing eyes) who had landed there in a flying saucer.

The incident was raised in parliament (and recorded in Hansard, lest you should doubt me) as an argument for banning American horror comics, which were clearly (it was said) the cause of this undesirable behaviour.

Going back a little further, the New Delhi story is strongly reminiscent of the late-Victorian London myth of Spring-Heeled Jack, a fire-breathing semi-humanoid (with wings) whose terrifying but harmless existence was attested to by countless witnesses of the most unimpeachable respectability. Jack went the opposite road from the Necropolis chimera, starting off as urban myth and ending up as a character in penny-dreadful comics, where he invariably turned out to be, like so many misunderstood villains of Victorian fiction, a wronged and dispossessed heir - albeit equipped with a splendidly engineered suit of wings and springs.

It doesn't do, you see, to think that delusional beliefs are the province of the ghost-ridden, pre-rationalist past, or of cultures allegedly less sophisticated than our own. Communal irrationality flourishes more vigorously in the West today than at any time since the witch-burning seventeenth century: how else can you explain the fact that millions of perfectly ordinary people have made feng shui - spiritual furniture-arranging - into a zillion-dollar business here and in the States?

Or that "alternative medicines" like reflexology, which amounts to foot-rubs as a medical discipline (A Toe for Every Woe; they can have that slogan for nothing, if they like), are now so respectable that you can get professional qualifications in them to distinguish yourself from practitioners who are only pretending to be real quacks? Or that investing pension funds in internet start-up stocks was wiser than the seventeenth-century Dutch practice of sinking one's life savings
into a handful of rare tulip bulbs?

For a long time - the forties to the seventies, inspired by psychiatrist's-couch cartoons in the New Yorker - the best-known image of delusional belief was the fellow who thought he was Napoleon. So much saner than he, are the rest of us today that when George W Bush says he believes he won an election to be president of the United States we all agree with him. But you'll have to excuse me now: the Monkey Man's in the foyer, and I've got an exclusive interview.

The Asian-based news agencies were much more sober in their reportage, as this item - again from the 23rd May from IndiaNews reveals:

MONKEY SEE, MONKEY DO, MONKEY WASN'T.

by Nirmal Ghosh.

NEW DELHI - The week-long monkey-man episode, now in its death throes, underscored the vulnerability of the human mind to rumour and misinformation - especially in matters that seem to be out of human control. The poorer areas where the monkey-man transformed the night have few telephones and probably just a handful of Internet connections. Yet the rumours flew through the narrow alleys and across ramshackle rooftops with cyberspace speed. Newspapers and television helped spread the scare as reporters told stories of alleged attacks by the half-simian, half-human creature. But nobody - at least in the early stages - examined the possibility of mass hysteria and willing self-delusion. By the time those options were considered, it was already too late. Four people died of injuries sustained in panic-stricken attempts to flee a probably imaginary assailant, and half the capital's police force was tied up in a frustratingly futile exercise. The areas where the hysteria took hold are low-income, comprising largely of immigrants from rural India, many of them beholden to superstitions.

Conditions can be dismal: no entertainment, no regular water or electricity supply, no proper sewerage. The monkey-man embodied much of what the human mind is afraid of: it could see in the dark, it possessed strength and agility, it did not leave footprints, it could change shape, it had red glowing eyes. In short, a potent cocktail of wild beast, supernatural being or plain lowly criminal, visiting people who had, by and large, not had much education beyond high school and are leading stress-filled lives on the margins of a big city. In the hot summer months, many sleep on rooftops. A few pegs of alcohol and a round of discussion about the monkey-man, and it is easy to understand the sort of nervousness that must have taken root.

The government, through the police, contributed to the situation with its slow response, although it may be said that the police could not rule out a "real" mischief-maker in the shape of the so-called monkey-man until it became obvious that it was hysteria, which was the issue.

The basis of the hysteria was fear of the unknown and the uncontrollable. Then the media made the phenomenon national news. It was information mutated into a wild and mischievous spirit accountable to no one.

The Straits Times - one of the oldest and most respected English-language Asian newspapers (based in Singapore) - provided the first attempts at a cultural analysis, placing the Monkey Man saga within a cultural framework of other urban panics. Again on the 23rd May they wrote:

- CFZ YEARBOOK 2004 -

20 May 2001 INDIA: Hysteria - The Indian experience

MUMBAI - 1960s, 1985-87 mass hysteria over unexplained killings ...

Hysteria - The Indian experience MUMBAI - 1960s, 1985-87 mass hysteria over unexplained killings gripped Mumbai twice. The first time, the killer was identified as Raman Raghav, who was finally arrested for the murder of 42 people. By 1985, the killings started again, but the mystery murderer remained elusive. Was it a man, stoneman, or something else entirely?

Whoever it was, he committed a series of 12 murders in the Sion and King Circle suburbs between 1985 and 1987. The killer, whose modus operandi was to crush the head of his victim with a huge stone, remains untraced. He is, perhaps, the most mysterious serial killer in Mumbai's crime history. The sleeping beggars' heads were smashed with an incredibly heavy stone, sometimes weighing as much as 30 kg. The stone killer never attacked beggars who slept in groups. In most cases, even the victims' names could not be ascertained since they either slept alone or had no relatives or friends who could identify them. The Matunga police were baffled by the killings. When they inspected the murder spot, they found no significant clues which led to the murderer. Although the police invariably found a huge stone near the spot, even the sniffer dogs brought in to track the killer lost his scent. Since most victims were beggars, robbery was ruled out as a motive.

The police then began a search for eye-witnesses to the killings. During the search, the police came across a waiter in an Irani restaurant near Sion Circle. The waiter was perhaps the only one who had managed to escape before the stone-man could crush his head. But he was unable to recollect the face of the killer. Shortly after his interrogation, the stone-man struck again. A rag picker was hacked to death near King's Circle. The murderer had begun to move out of the Sion area and towards Matunga. Panic-stricken residents from Matunga urged the local police to intensify their vigilance during the night. The case was marked high priority - with more than 100 persons patrolling the area at night. People were warned not to sleep alone on the footpaths. After the twelfth murder in 1987, the killings stopped abruptly.

Which still leaves the police wondering whether the psychopath will one day strike again....

KOLKATA, 1989. In the history of the nation's oldest detective department, stone-man remains one of the greatest mysteries unsolved. Memories of stone-man still haunt the sleuths who investigated the case. He first struck in mid-June, 1989. His victims - all pavement dwellers. His murder weapon - a stone or concrete slab weighing over 14-15 kg, which was left near the victim. The toll - 13 in a space of six months. The first incident was reported from the Laldighi area, a stone's throw away from the police headquarters at Lalbazar. Circumstantial evidence indicated that the assailant was tall and well built. "He used to drop the concrete slab on the sleeping victim's head, crushing it completely. There was no hammering involved," said a senior detective department official. As more killings were reported, panic set in. "Every month, one to three murders were reported," recalled Prasun Mukherjee, IG (law and order), who was then the detective department chief. "Nobody saw him as he carefully chose only those who slept alone. The victim was given no chance to raise an alarm," he added. Soon rumours of sightings began to do the rounds. While some described him as a giant, others recounted seeing an apeman prowling the streets. Many saw it as the handiwork of an evil force.

"We picked up some rumour-mongers. However, none of them could give a proper description for our artist to draw his picture," a senior detective said. The department tried to identify a pattern in the killings. Initially, they thought the murderer struck only on Fridays. "However, we were proved wrong. Also, the murders were neither age nor gender specific," he recalled. From psychiatrists to tantriks, every possible angle was investigated by the police to work out a motive. According to one theory, it could be someone acting on a tantrik prescription, to sacrifice a certain number of human lives to attain spiritual fulfilment. The murders were limited to a few police station areas around central Kolkata. "We were in such a state of panic that our officers would leave early evening and start removing concrete slabs from areas where pavement dwellers usually slept. We also appealed to pavement dwellers to sleep together in clusters as there was safety in numbers," a police officer said. The only positive fallout of the whole affair was the decline in nocturnal crime. "The killings went on till early March 1990.

We arrested a man but he turned out to be insane. We could not conclusively prove that he was the stone-man, but after the arrest, the murders stopped," said Mukherjee.

BANGALORE, 1996. House after house had the Kannada inscription Naale baa (Come Tomorrow) on its front door. For months together in 1996, the ghost of an old lady would coming knocking at your door. Open it and you invited certain death, not just yours but others in the family too. What started in the slums of the Srirampuram area, spread rapidly to the western and northern parts of the city. Soon enough, not just slum-dwellers but even middle class homes in Vijayanagar, Rajajinagar and Malleswaram were putting up the signs like name-plates on their main doors. "It all began in the Tamil-speaking slums of Srirampuram when two or three households lost family members without any apparent rhyme or reason. Then the ghost theory came up.

The slum dwellers put up `Naale Vaa' in Tamil. Soon, residents in other areas got to hear of the ghost and were taking precautionary measures," a Malleswaram old-timer recounts. The ghost was bad enough to kill, but kind enough to spare you if you requested it to come the next day. "With the inscription up permanently, the devil would never come in. You literally bought an additional day of life every day," recalls a resident of Vijayanagar. The ghost gave up one day. Bangaloreans had got clever....

AND 1998 `You've been injected with the HIV virus. Welcome to the world of AIDS,' the sticker supposedly proclaimed.

In 1998, the murmurs began in Bangalore, went on to become a cacophony which scared, irritated and troubled the citizens and then died away. Nobody could quite disprove it. But then, nobody proved it was true in the first place either. Film-goers. Local bus commuters. Solitary walkers. Sinister messages, mostly left behind in the form of stickers on one's jacket or the bike. The rumours spread thick and fast. People being jabbed with syringes containing the HIV virus in dark theatres - on their way back from work and in crowded supermarkets. Nothing original, though. A similar scare had taken place a few days before at a cinema hall in Chennai. Were they pranksters or was it a lunatic? Bangalore went into a tizzy for over a month. The counter-measures? Look over your shoulder. Avoid late-night film shows and never ever give strangers a lift. The police said that none of the 80-plus police stations in the Bangalore city limits had seen a complaint registered.

A senior police officer had in fact told The Times of India then that the police had also heard of several such cases, but nobody had formally approached the police. The stories got more outlandish with every passing day. Callers to newspaper offices related tales of how innocent victims were dying of AIDS because of the injections. Little children in front of schools, old men taking a walk... And then just as suddenly, the rumours died a quiet death. Rumour or truth, one still hasn't been able to say. But Bangaloreans did reel under the viral attack for some time. (With inputs from STOI Mumbai, Krishnendu Bandopadhyay in Kolkata, Sriranjan Chaudhuri in Bangalore).

Three days later, the Indian press discovered what their South and Central American counterparts had known for years - if you have a paranormal or fortean news story which is selling newspapers, the best way to sell more newspapers is to incorporate the buzzword which has proved to be a success into a news item which is only tangentially related. I first noticed this in Mexico in 1998, when I was investigating the chupacabra in the trip, which later became my book Only Fools and Goatsuckers. Although there had been a long tradition of Vampiric attacks on both humans and livestock in Mexico, it wasn't until the mid-1990s and the global proliferation of the chupacabra ethos, and even more importantly the global proliferation of the well-known image of the chupacabra - a weird kangaroo-like beast with spines up its back - that anyone even started linking the two phenomena. As I write in my new book Monster Hunter:

"Indeed, even the earliest so-called chupacabra attacks had been blamed on a mysterious winged big cat, and it was not until after the events of 1995 when the - by now - ubiquitous image of the kangaroo like creature from Puerto Rico was splashed across the world's media that anybody (mostly self-styled UFO experts), started to draw the inference that the two sets of incidents were even slightly related."

In this case the connection between the Monkey Man panic and the new fortean phenomenon - a supposed lycanthrope in Assam seems even more tangential as a quick perusal of this story from the Indian Express News Service on May 26th shows:

MONKEY-MAN MORPHS INTO WEREWOLF IN ASSAM VILLAGES

Guwahati, May 26, 2001:

AFTER Delhi, it is now the turn of Nalbari district in Lower Assam where a strange wolf-like creature is reportedly out to create panic. Going by reports pouring in to the state capital, over a dozen people have been already attacked by the mysterious wolf-like creature, compelling people to organise all-night watch groups to protect themselves.

The creature has been reportedly sighted in villages in the Nathkuchi and Tihu area in the backward district about 90 kms west of Guwahati, with people complaining that it can enter houses even if doors are closed. People say it roams about, often becoming invisible before attacking them, said Prafulla Kalita, a government employee in Nalbari town, who hails from a village in the Tihu-Nathkuchi area. Kalita, however, has not seen the creature herself.

One Assamese daily quoted villagers as saying that the creature has a furry body, looks like a bear and disappears when rays of light are directed toward it. The villagers are spending sleepless nights and the situation has become even worse with erratic power supply, especially after sunset. When contacted, Nalbari Deputy Commissioner B. Kalyan Chakravarty dismissed the reports as nothing but figments of people's imagination resulting in mass hysteria.

Chakravarty also attributed this to the widely-published reports of the mysterious monkey-man in New Delhi as another reason behind the panic in the Nalbari villages. We sent out teams of officers, including the police, to these villages, and after careful examination found that it is just some kind of panic created following reports of the monkey-man in Delhi. Such stories always create sensation among the villagers, who easily give in to superstitions and ghost stories, he asserted.

Panic-stricken villagers have even approached the Army engaged in counter-insurgency operations in the sensitive district. "We sent out men to verify the incidents, but till date not one person has been able to substantiate his claim that he had seen a mysterious creature", said Col Sanjeev Sinha, CO of the 5 J-K Light Infantry posted at Nalbari.

Interestingly, the Assam Science Society, an apex body of scientists and science propagators in the state, has come out with a detailed report, which dismisses the existence or appearance of any mysterious creature in the district.

The report prepared by a five-member team headed by Dr Karuna Kanta Patgiri, president of the Bajali branch of the Society, said all the 16 persons it interviewed had reported that they were in sleep when the attacks occurred. The people said they first heard a noise on the tin roof of their houses, following which they felt that something was trying to clutch them with sharp nails. But every person admitted they were half-asleep when they experienced this, the report said.

The story was picked up by Ananova who morphed it slightly and - the same day - promulgated it onto the international stage:

First 'Monkey Man' - now villagers panic over 'Bear Man'

As Monkey Man hysteria dies down in Delhi, villagers in north-east India are claiming a new menace is on the prowl - Bear Man.

Villagers in areas of Assam claim the creature makes itself invisible before attacking people. Bear Man apparently disappears when caught in a ray of light.

More than a dozen people say they have been attacked by Bear Man.

Police have dismissed the sightings and reported attacks as figments of the imagination.

Deputy Commissioner Kalyan Chakravarty says the panic may also have been fuelled by reports of Delhi's Monkey Man.

He says: "Such stories always create sensation among the villagers, who easily give in to superstition and ghost stories."

The creature has reportedly been sighted in the Nalbari district. Villagers have organised all-night watch groups to protect themselves, says the Indian Express.

This report also showed the panic spreading out of India to other parts of the sub-continent because - as Loren Coleman pointed out at the time - the District of Moulvibazar is in Bangladesh. The story was also carried on the same day by the Indian Daily Star who treated it in a far more sober fashion than their British counterpart would have done:

STRANGE ANIMAL CREATES PANIC IN MOULVIBAZAR.

Our Correspondent, Moulvibazar:

Three women including a girl child were injured by a strange animal in some villages under Prithim Pasa union of Kulaura upazila under Moulvibazar district.

A panic situation was created in the area.

According to the local people, the odd-shaped strange animal entered into the house of Basit member of Rautgaon village at night recently. Wife of Basit member was injured by the animal's scratch. On the following night, the animal went to a house of village

Kazirgaon where another woman received injury. A girl child aged about 12 to 13 was injured by the same animal at a house of Uttar bazar of Rabir bazar.

Upazila Nirbahi Officer (UNO) of Kulaura was not available for comments when this correspondent tried to contact with him over telephone. A Sub-Inspector (SI) of Kulaura Police Station said over telephone that they did not receive any information about such incident.

People of Prithim Pasa informed that the aforesaid animal looks like a monkey, but its face looks like a man and its height is 3 feet.

However, villagers of the area were patrolling the villages at night. Most of the women are passing night in pucca houses of their neighbours. The matter should be investigated, opinioned the local elderly people.
However, this wasn't the first such usage of the Monkey Man ethos to sell newspapers. The previous day, The Times of India had used the monkey man "brand name" to sell a particularly uninteresting story about snakes, to which they had added some vaguely believable fortean motifs to make the story more palatable:

AFTER THE MONKEYMAN, HERE'S MR. HORROR.

by Bindu Jacob.

NEW DELHI, AUG. 16. Delhiites have barely managed a breather from the monkey-man and they find their hands full yet again. And this time round it is no iron-clawed creature that is spelling terror in the stillness of night. The fear comes from altogether different quarters: crawling snakes.

Following the death and a near-death experience of two residents of a slum cluster in Punjabi Bagh here earlier this month, the ground has been laid for horror tales to unfold. It all began about a fortnight ago when 18-year-old Rakesh was bitten allegedly by a snake and died hours later. "Rakesh had come along with his family to Delhi in search of a job. He was sleeping in his uncle's jhuggi when he was bitten by a snake one morning. His family rushed him to a nearby clinic, but he died later while being shifted to a hospital. His family has since gone back to his village," says Lal Babu, the area Pradhan.

A few days later, 16-year-old Anil was bitten while he was asleep in his house. "I woke up screaming with pain and both me and my brothers saw a green and white snake escape into a hole at the corner of the house. We cut open the bite area, sucked out the blood, and rushed to hospital. The hole in the house has been cemented since," claims Anil.

And the last such "incident" was reported this past Monday morning. A woman was hit on the hand by some "object" while sweeping the floor. Though flustered, she proceeded with her household work. Shortly afterwards, she felt dizzy and fell down. Since she did not suffer any external injury, the "ghost theory" has now replaced the "snake scare" among the 300-odd residents.

Since that incident, the area has been so tense that all injuries are invariably passed off as snakebites. Last year, too, around the same time panic had gripped the area after the residents reported sighting of two snakes. Subsequently, a snake charmer was summoned to ensnare the snakes. Rumours are now flying thick and fast and the fear of the residents is almost taking the form of mass hysteria. One tale doing the rounds is that the snake even assumes the form of a human being and can be sighted on rooftops. Another is woven round the death of Anil's father. Residents claim that it is his wandering "soul" which is unleashing terror. Of course no one is ready to buy the possibility that it is waterlogging in nearby marshes due to the monsoon which is forcing the snakes to come out and attack whoever is around in the thickly populated slum clusters.

The original Monkey Man episodes seemed to die down soon after. I believe that the reason for the graphic difference in reporting styles between the South American reports of mutilated livestock, and the Indian Monkey Man stories are twofold. Firstly, the political situation in India was far more stable. Although the ongoing problems with Pakistan were still there, in the months prior to September 11th, following which the hunt for Osama Bin Laden, and the war in Afghanistan brought racial tensions between Hindus and Muslims to the surface, against a relatively stable political background it was possible to analyse these news items by using a socio-cultural model. Secondly, in a stark and telling contrast to the events in South America, there was no actual physical evidence.

Then, on September 11th, 2001, the world changed forever. When the next series of Monkey Man attacks took place in the spring of 2002, the difference in tone of the reporting - and also the difference in what was actually reported is enormous.

On the 9th July, an Indian news agency filed the following story:

Police hunt 'monster man'

People are fleeing Indian villages to escape a man who allegedly plucks flesh from his victims with long claws.

Around 50 villages in the Ghazipur district of Uttar Pradesh are in fear of the so-called monster man who is reported to have attacked 36 people.

One of the man's victims claims to have seen red and green lights coming from his body before he struck.

He is also rumoured to have long hair, blood-shot eyes and claws like a tiger.

Weddings have been cancelled and people have stopped sleeping outside their homes at night despite the summer heat, reports Sify News. People have fled Loharpur, Balua Tarao, Dharam Purwa, Jagatpur and many other villages.

Director general of Police RK Pandit has launched an investigation into the alleged attacks by the man, known as Mooh Nochva.

Inspector General of Police in the Varanasi zone, S. N. Singh, said a team of officers is trying to identify the man.

Sify reports his victims have included a woman called Pikkhi who had flesh ripped off her face, neck and arms, and student Jai Prakash Paswan whose face and arms were badly scratched.

On the same day, this story appeared on Ananova:

CHILD DIES AS 'MONKEY MAN' SPARKS PANIC

PATNA, India:

A child was killed in a stampede by panic-stricken residents of an eastern Indian village trying to escape an elusive monkey-like creature, which has been terrorising them for days, media reports said yesterday.

The six-year-old boy was fatally injured on Monday when he fell off a roof during the panic, sparked by reports that the "monkey man" was near, the reports said.

A woman fractured her arm in the stampede, which occurred in Lahlan village, 90km from Bihar state's capital Patna.

Police could not confirm the incident but said they had received numerous reports from various parts of the state of a mysterious "ape-like creature" attacking people, a year after a similar scare gripped New Delhi.

Some alleged the Bihar creature "jumps and sparkles with red and blue lights," while others described it as resembling a machine, operated by remote control.

Residents of Khupri village, 40km west of Patna, reported sighting the "monkey-man" in the form of "a shining object" falling from the sky, a police spokesman said.

The local police have issued a stern warning to people not to spread rumours about the "monkey-man."

"We will take strict action against anyone trying to spread rumours and causing panic among the people," the police spokesman said. Ananova Monkey man scare grips eastern India

The tone was infectious. The French Agence France-Presse announced:

Patna, July 21

Almost a year after residents of Delhi were terrorised by the alleged attacks of an elusive monkey-like creature, the eastern city of Patna says it too is facing similar attacks from an ape-like animal.

Local newspapers here are full of reports of "mysterious monkey attacks" but police have scotched the stories and warned of strict action against those spreading rumours.

"There are rumours of a monkey-like machine, referred to as monkey-man that attacks those sleeping on rooftops and in open places at night," ON Bhaskar, Patna's police chief said.

"But it is a pure rumour as no one has actually lodged a case in any police station. There has not been any recognized case of injury. We have warned the public at large to be on guard against any rumour and help cops arrest those who spread such rumours."

Residents of Patna however were not reassured, with local newspapers headlining quotes from eyewitnesses and victims of the "monkey man".

"The monkey-man attacked and injured my son-in-law Joginder Singh Friday night when he was sleeping on the rooftop of my house," a newspaper quoted Bhagwat Sharan Singh of Patna's Mainpura colony as saying.

"He (the attacker) looked like a monkey." Some even alleged the creature "jumps and sparkles red and blue lights". Others described it as resembling a machine, operated by a remote control and "handled by anti-social elements to terrorise people".

Police chief Bhaskar said hospitals and doctors had been asked to report any case of injury attributed to the mysterious creature. Meanwhile, fear of the "monkey man" took an ugly turn when a group of people beat up a Hindu Sadhu (saint) with a flowing beard, on suspicion that he was the creature.

The incidents reported in Patna seemed almost identical to the attacks reported in the Indian capital last May, when for more than a month, a "mysterious monkey-like creature" besieged New Delhi.

Descriptions agreed the creature was "black" and "ape-like" with "sharp claws," but varied on its height, with some reporting it was over six feet tall with red eyes, while others said it was about two feet tall.

The same day, The Hindustan Times broadcast a slightly more sober (but still horrifically sensational) account of the events in Uttar Pradesh. Reading the article, which follows it, is hard not to draw inferences between this parochial hunt, and the enormous manhunt, which was taking place in the Bora Bora, caves only a few thousand miles to the west.

'MONSTER WITH CLAWS' TERRORISE UP VILLAGES

by Vinay Krishna Rastogi in Lucknow

Villagers in Uttar Pradesh have been terrorised by a man who allegedly plucks flesh from the mouth and other body parts of his victim with his long and deadly claws. The "monster-man" has attacked 36 people so far.

At least 50 villages of Ghazipur district are in grip of "fear of the devil", who they call "Mooh Nochva", and at every passing day, the terror is spreading to more and more areas.

Director general of Police RK Pandit has directed the IG to rush policemen to protect villagers and to ascertain the identity of the dreaded man playing the role of a monster.

A married woman, Pikkhi, who arrived in Balua Tarao village to stay with her maternal uncle Treveni Rai was attacked by the mysterious monster. In the attack, he ripped flesh off her face, neck and arms. Kedar Dushad of the same village was also attacked

Among the three dozens victims of the mystery man is an Intermediate College student of Balua tarao village Jai Prakash Paswan. He claims he saw lights of red and green colour emitting from the body of the mystery man before he attacked him.

His face and arms were badly scratched.

Many marriages in the villages have been cancelled due to the fear of the man "considered by the district administration as a probable psychopath".

The area is awash with rumours. There are several versions about the man having long hair, blood-shot eyes and claws like that of a tiger.

Yet others absurdly describe him as a robot that emits light when attacking his victim.

Inspector General of Police, Varanasi zone S N Singh said on phone that a team of sleuths has been dispatched to ascertain the identity of the man.

Earlier, it was believed that the attacker was a woman, but now villagers are certain that the person is a weird, strange-looking man

People have fled Loharpur, Balua Tarao, Dharam Purwa, Jagatpur and many other villages as the mysterious man might attack any time if the victim is alone. He attacks and disappears in a flash.

Inhabitants of over 50 villages have stopped sleeping outside their homes at night in view of the blistering summer season.

By the 11th August the reports had become so bizarre that the following story from The Times of India sounds so like something out of one of the more unpleasant and less believable stories by H.P. Lovecraft that it almost beggars belief:

A team of experts is being sent to investigate mysterious attacks in India. Villagers in Uttar Pradesh claim to have been attacked by a creature described as resembling a flying octopus, which targets people's faces.

Now the government, keen to play down rumours about the mysterious creature, has asked experts from the Indian Institute of Technology in Kanpur to investigate.

The team will examine the role light plays in the attacks, as some of the incidents took place at night, reports The Times of India.

Government minister Dipti S Vilas said it had not been proved the incidents were caused by the creature, known as muhnochwa.

He cited a case in Hardoi where a woman claimed to have suffered scratches on her face after being attacked outside her house. Medics said the scratches were not caused by muhnochwa and she later admitted getting the scratches after falling inside her house.

He also said an 11-year-old boy who died in Jaunpur had not been killed by muhnochwa, although the cause of death was unknown.

Mr Vilas said if the rumours didn't stop, devious people might take advantage of the situation.

It is tempting to theorise that this report - and in particular the reference to "devious people" (which so very comfortably mirrors the reports of "dark forces" which surfaced the following year in the UK during both the David Kelly and Paul Burrell affairs) - are a snide blow at a story which had appeared in the New York Times on the previous day. The story could almost have been a piece of journalistic identikit, mirroring closely - as it did - both the reportage of the 2001 South American cattle mutilation wave, and much of the journalistic take on both Puerto Rican and Mexican "Chupacabra" episodes:

VILLAGERS BLAME UFO FOR INDIA ATTACKS

SHANWA, India (AP) - It comes in the night, a flying sphere emitting red and blue lights that attacks villagers in this poor region, extensively burning those victims it does not kill.

At least that's what panic-stricken villagers say. At least seven people have died of unexplained injuries in the past week in Uttar Pradesh state.

"A mysterious flying object attacked him in the night," Raghuraj Pal said of his neighbor, Ramji Pal, who died recently in Shanwa. ``His stomach was ripped open. He died two days later."

Many others have suffered scratches and surface wounds, which they say were inflicted while they slept. In the village of Darra, 53-year-old Kalawati said she was attacked last week and displayed blisters on her blackened forearms.

"It was like a big soccer ball with sparkling lights," said Kalawati, who uses only one name. "It burned my skin. I can't sleep because of pain," she said.

Doctors dismiss the stories as mass hysteria.

"More often than not the victims have unconsciously inflicted the symptoms themselves," said Narrotam Lal, a doctor at King George's Medical College in Lucknow, the state capital.

The police have another explanation: bugs.

"It is a three-and-a-half-inch-long winged insect" that leaves rashes and superficial wounds, Kavindra P. Singh, a superintendent of police, told the Press Trust of India news agency. Police drew this conclusion after residents of one village found insects they had never seen before.

Villagers are unconvinced. In the most affected area, the Mirzapur district, 440 miles southeast of New Delhi, people have stopped sleeping outdoors despite the sweltering heat and frequent power outages.

Villagers also have formed protection squads that patrol Shanwa, beating drums and shouting slogans such as, "Everyone alert. Attackers beware."

Some accuse district officials of inaction and failing to capture the "aliens." One person died Thursday in nearby Sitapur when police fired shots to disperse a 10,000-strong crowd demanding that authorities capture the mysterious attackers.

"People just block the roads and attack the police for inaction each time there's a death or injury," said Amrit Abhijat, Mirzapur's district magistrate, who claims he has captured the UFO on film.

MANJARI MISHRA

This farrago of sensationalist bilge is - I believe - exactly what the original commentators on the subject who were quoted at the beginning of this paper wanted to avoid. The New York Times article - in particular - portrayed India as a nation of backward peasants. This is a long way from the technological 21st Century nation which the Indian government - and the media - are desperately trying to sell to the rest of the world. It seems unlikely - you can imagine the pundits and politicians in New Delhi thinking - that British Telecom (or any of the other major companies who have recently relocated their call centres to the sub-continent), would want to invest in a nation whose populace believed that they were under imminent peril of being attacked by an unknown flying Stingray with metallic claws.

It was about time for politicians to get involved. Deciding that it was far too late for a damage limitation exercise, as the Times of India reported on 21st August, Indian politicians decided to do what politicians do best and to blame the Monkey Man hysteria on the real enemy - the party in opposition:

LUCKNOW: The raging muhnochwa (face-scratcher) controversy dogging Uttar Pradesh for the last two months took an unexpected twist on Tuesday with chief minister Mayawati blaming the opposition Samajwadi Party for raising the monster bogey in order to destabilise her government.

Mayawati on Monday lambasted her political rivals for engineering violence in Barabanki that claimed the life of one of the villagers agitating against the recurrent attacks by the mysterious object.

And even as debate over such a possibility rages, the basic question still remains unanswered: What exactly is muhnochwa which has sidelined and even overshadowed all other activities in the state?

The issue led to unprecedented bedlam in the state legislative assembly on Friday as Samajwadi MLAs, alleging a foreign hand in the attacks, demanded a thorough inquiry.

It has been probed into by the intelligence agencies, forensic experts, medico-legal experts, electronic enjoiners, physicists, psychiatrists, IIT faculty members, local administrative and police officials - but each version has only confounded the prevailing confusion.

The bizarre happening, which led to much panic, began in Mirzapur, Chandauli and Varanasi and gradually spread over the adjacent districts of Sultanpur, Sitapur, Rai Bareily, Pratapgarh, Lakhimpur Kheri, Pilibhit, Lucknow, Kanpur, Allahabad... all of eastern Uttar Pradesh.

And as the death toll of the victims rises by the day, agencies pitted against the unknown enemy find the pressure building up from all quarters.

The harassed Lucknow police are even contemplating booking rumour-mongers under the National Security Act. Forest officials are worried over the merciless killing of squirrels, foxes, bats, hyenas and wolves by terrified villagers.

Predictably, the incidents have contributed to many a goof-up. Like a high-ranking police official stating it was a "technologically developed insect... released by anti-social elements". Or the much-hyped carcass in Lucknow that turned out to be a harmless insect from the grasshopper family.

Even IIT scientist became the butt of ridicule when they propounded the theory that muhnochwa was a "ring of gas" or "ball of lightning" emerging due to the drought.

But while some injuries have been found to be self-inflicted, and there have been stray incidents of mischief, the "genuine cases" have baffled the authorities.

There has been no mistaking the bloody scratches on the victims' bodies, some of which have suffered severe trauma, requiring hospitalisation and prolonged treatment.

The victims' descriptions are eerily similar - blue and red lights, a moving object, sudden attack and electric shock - defying all rational explanations.

So, when an expert in Allahabad demanded the intervention of the National Human Rights Commission, only a few in the city found it funny.

The official explanation was now clearly that the Monkey Man, the flying Stingray, the giant robot Monster, and all the other quasi-fortean happenings of the last few years were also down to the misidentifications of one beast - a rare type of Grasshopper. Surprisingly, nobody apart from me and the other guys at the CFZ found this explanation even slightly amusing. Even more surprisingly - to the best of my knowledge - no-one, until now, has drawn a parallel between this ludicrous explanation from the Indian government, and the equally ludicrous one which surfaced in 2001 from the beleaguered Government of Argentina. The official culprit in the government investigation into the animal mutilation wave which had terrorised the pampas for the best part of the year was a - almost exclusively vegetarian - species of fieldmouse. It is my heartfelt - if rather cynical - belief that the main job of government press releases, especially in the face of a crisis, is to try and deflect even larger amounts of social panic than they already have. In the late lamented BBC comedy series Yes Minister, Sir Humphrey Appleby - a senior British civil servant - is quoted as saying that one of the major mechanisms of government is as follows:

- CFZ YEARBOOK 2004 -

1. Something needs to be done

2. This (no matter how ludicrous or ineffectual) is something

3. Let's do it

In the face of the evidence from Argentina and India, and in particularly the evidence of history in that, in the wake of these - admittedly ludicrous - government statements (which can only be seen as a panacea to the lumpen proletariat), that in the end, the panic dissipated, and the much-feared breakdown of law-and-order singularly failed to happen.

The Government statement was hidden in this article dated 20th August from the Times of India:

LUCKNOW: The hype generated around the 'scratch monster' (muhnochwa) caught in Lakhimpur Kheri last Friday and brought to the Lucknow University for identification ended in a whimper on Monday, when senior professors of LU zoology department described it as a harmless insect belonging to the grasshopper family.

Head of the zoology department, Prof KC Pandey, told mediapersons that the three-inch long insect was a rare species and was mostly found on banks of the river. Its biological name is Schizodatylus Monstruosus.

The insect has six legs and mostly feeds on small insects. Zoology department experts denied that the insect could in any way harm human beings.

The legs have small nails which can only make a slight mark on the skin and not at all the kind of scratches being reported by people, who claim to be victims of muhnochwa, they stated.

Terming the muhnochwa scare as a rumour, Prof. Pandey said the insect brought for identification did not emit any lights.

The professors conducted a detailed study while analysing the insect. They consulted various books and references before going public with their observations.

I think that the most important conclusion, which can be drawn from this data, is not that quasi-fortean episodes of what may - or may not - be Zooform Phenomena are inextricably linked with the social and political background of the country. I feel that I have proved this on a number of occasions, and that it would be a waste of breath - and paper - for me to cite the Monkey Man episodes as yet more supporting evidence of my thesis. To me, the most important thing about this event in India during the summer of 2001 and 2002 is how the Government reacted. The Paranormal press across the world has repeatedly accused governments of all nations of not taking Fortean phenomena seriously. I believe that the evidence presented above actually lets the aforesaid governments off the hook. I believe that governments all over the world - and whether the conspiracy theorists like it or not, although there is no new world order, most government leaders across the world were educated at one of only half-a-dozen establishments in the UK or USA, and therefore have learnt their craft through roughly similar channels - are perfectly aware of the reality of fortean phenomena - at least in the socio-cultural sense. They are not above using such phenomena - or the belief in such phenomena - as a whitewash or smokescreen for their own covert activities, and when - as in the case of the Dehli Monkey Man - a situation arises which is potentially beyond their control, they have a perfect weapon to use against it. Citing a ludicrous explanation like field mice or grasshoppers is not - I feel - that are far from the advice of my mentor Tony Shiels.

He once told me that the only way to combat Psychic backlash - the inevitable wave of misfortune which befalls people who become too closely involved in the more grotesque Zooform Phenomena - is with laughter and a sense of the absurd. Possibly governments across the world have been taking lessons from him.

(Editor's note: apart from obvious spelling mistakes, we have not changed grammatical or language idiosyncrasies within the news reports. Indeed, we believe that when a luckless journalist complains that a senior police officer was not available when "this correspondent tried to contact with him over telephone", it is a plaintive - and almost poetic - complaint which would be far less well expressed in the more conventional uses of the English language.)

THE CRYPTOZOOLOGY OF DR WHO
by Richard Freeman

One of the questions that I am most often asked is - what inspired me to become a cryptozoologist? I always reply in two words - Doctor Who!

For those of us of a certain age, Dr Who was an integral part of our lives, as much a part of what made Britain British as fish and chips, or the white cliffs of Dover. For half-an-hour every Saturday, we would be transported away from the mundane, lackaday world of everyday existence, to bizarre and disturbing vistas of cosmic strangeness.

Dr Who was unrivalled in its sheer weirdness. In what other programme could you have a Buddhist sect summoning up interdimensional, giant psychic spiders, or swarms of 4 ft long mutant maggots bursting out of slag heaps in Wales? In these modern days of banal television and the nanny state, it is hard to conceive of any teatime programme involving giant rats gnawing off people's legs, opium addicts, prostitutes, and a homicidal ventriloquist's dummy with the brain of a pig. In fact all these things were included in just one Doctor Who adventure - The Talons of Weng Chiang (1977). One cannot conceive of dreary and unimaginative programmes like Star Trek, and its increasingly vacuous spin-offs, having story lines involving animated shop dummies that burst through shop windows and attack people, lumbering robot mummies that crush people to death, or swarms of super-intelligent aquatic reptiles rising from the sea to menace mankind.

Inversely, during its 26 year run, Doctor Who has on several occasions, been influenced by cryptozoology. In this article, I will examine the Cryptozoological, or cryptozoologically inspired, creatures which have appeared in this - the greatest of science-fiction television shows.

The first Cryptozoological inspired story was in season five (1968). The Abominable Snowmen starred Patrick Troughton as the second Doctor. In this story, the Doctor and his assistants - Jamie and Victoria - travel to Tibet in 1935. They meet up with an English scientist called Professor Travers who is searching for the legendary Yeti. The Doctor is returning a sacred bell called the Ghanto to the Detern Monastery. He discovers that the normally shy and peaceful Yeti have turned into savage killers. It transpires that the yeti are in fact robots controlled by a quasi-Lovecraftian Cosmic entity known as `The Great Intelligence`. The intelligence has possessed the Doctor's old friend Padmasambhava, Abbot of the monastery. It controls the robot Yeti by metallic spheres in their chests, and manifests itself in a remote cave as a pool of ectoplasm like substance.

The Yetis themselves are hulking creatures with featureless faces. They are entirely covered in hair, save for their clawed hands and feet. They have rounded heads, no necks, and bizarre looking broad thighs. At the end of the adventure, Travers, the doctor and his companions see a real Yeti. It is depicted as being taller and slimmer than the Robot Yetis and it has an ape like face.

The sequel to The Abominable Snowmen - which was broadcast later in the season, was a story called The Web of Fear. Set 30 years after The Abominable Snowmen, it

- CFZ YEARBOOK 2004 -

A robot Yeti stalks the London underground in the 'The Web of Fear'

begins when a robot Yeti bought to London by Professor Travers is reactivated. This time around, the great intelligence manifests itself as a sinister web-like substance, spreading through the tunnels of the London Underground. The yeti are portrayed this time with glowing eyes, stalking through the unlit tunnels striking people down with their claws, while the Web envelops its victims. You wouldn't find anything like that in Buffy the sodding vampire slayer!

Thankfully, the real Yeti does not seem to be so dangerous. The name Yeti is thought to be derived from the Sino-Tibetan Metoh-Kangei meaning 'rock animal'. The other popular name for this creature comes from a bizarre mistranslation of Metoh-Kangei. An Indian telegraph operator sending details of one of the early expeditions back to London mis-spelt this as Metch-Kangmi, which means Abominable Snowman. The Yeti is not abominable, does not live in snow, and is not a man. It is a generally shy animal that lives in semi-tropical forests and is almost certainly an undiscovered species of giant ape. Another myth that the name Abominable Snowman perpetuates is that the creature is white. There has never been a report of a white Yeti. Their fur can be ginger, chestnut brown, or black.

There appear to be three different kinds of Yeti. The largest is called Dzu Teh and it can be three metres (10 feet) tall. It walks on two legs - like a man - and is it is immensely muscular, with a prominent sagittal crest on its head. This type of yeti is likely to be a surviving form of Gigantopithecus - a giant ape from the Pleistocene epoch. Its fossil remains consisting of teeth and jaw fragments have been found in China, India, and Vietnam. The structure of its jaw suggests that it fed mainly on bamboo and that it was a biped. It was believed to have become extinct 500,000 years ago. Many animals that lived alongside its - such as pandas, rhinos, elephants, and tapirs - are still around today. There seems no reason why Gigantopithecus should have become extinct.

Another type of Yeti - known as the Mi-teh or the Ye- to teh - is about as tall as a human being, although much more muscular. It walks both erect and on all fours. It may be a mainland species of orang utan. Orang utans today live only in Borneo and Sumatra, but in the Pleistocene, a larger form lived on mainland Asia.

The third type of Yeti is called Teh-lma and is about one-and-a-half metres tall. It constantly walks erect and may be closely related to the Orang Pendek of Sumatra

Travers is heavily based on the British explorer and cryptozoologist Peter Byrne. In 1959 he examined a mummified yeti's hand at the Pangboche monastery in Nepal. The hand was a holy relic kept by the Buddhist monks. Byrne wanted to take the hand back to the West to be examined, but the monks would not let him. So he stole the thumb and phalanx, replacing them with human finger bones so that the monks would be none the wiser. He gave them to his friend - the actor Jimmy Stewart - who was on a holiday in Nepal at that time. Jimmy smuggled them back to England where they were examined by the primatologist Dr W.C.Osman-Hill. He declared them to be larger than those of a mountain gorilla, and from some gigantic, unknown anthropoid. Unfortunately, the bones have since gone missing. What nobody knows is if they have been lost or stolen. The monastery itself was recently burned to the ground.

In season seven (1970), the third doctor - played by Jon Pertwee - encounters a race of highly intelligent lizard-men called The Silurians. These creatures are hibernating in some caves under a moor in Derbyshire. The Silurians evolved some time during the age of the dinosaurs, and were once rulers of the Earth. The entire race went into worldwide hibernation using stasis machines in order to avoid a massive meteorite impact on the earth during the Late Cretaceous period.

A scientist has stumbled across them, and is draining energy from a local nuclear power plant in order to awaken them. He hopes to use their scientific knowledge, but the Silurians are disgusted that the planet has become overrun with mammals. They release a Tyrannosaurus Rex into the cave system, and infect humans with a virulent disease that quickly spreads to London.

The Doctor finds a cure for this disease, and tries to convince both humans and Silurians to share the planet. Whilst the doctor is trying to convince the world governments to live in peace with the Silurians, they return temporarily to hibernation. Behind the Doctor's back, the Brigadier orders the caves to be blown up, sealing the Silurians in the forever.

The Doctor confronts a Sea Devil

In the season nine (1972) story, The Sea Devils, the Doctor encounters the Silurian's marine cousins. After oil drilling in the North Sea awakens them, the Sea Devils start attacking boats and oil rigs. Incarcerated on a nearby island in a fort, is the Doctor's arch-enemy - a rogue time-lord called The Master. The Master has contacted the Sea-Devils and is trying to stir up a World War. Once again, the Doctor tries to act as peacemaker between the two species, but the intervention of the master, and human xenophobia, thwart his plans.

These two stories pre-empt - and may even have been inspirational - to the theories of Dr Dale Russell of the Canadian Museum of Nature in Ottawa, Canada. Dr Russell postulated that if non-avian dinosaurs had not become extinct 65 million years ago, they would have now become far more intelligent than human beings. Using a small carnivorous dinosaur called Sauronitholestes as a model, he came up with a creature he called a dinosaurid. This was a humanoid reptile, with forward-facing eyes, and opposable thumbs.

The idea that evolution is constantly striving for great intelligence is a Victorian theory, largely abandoned. In fact, evolution selects the best traits to help animals survive in any given environment. This does not necessarily mean a greater intelligence. For example, as sauropod dinosaurs developed longer necks, their brains became smaller. This is because their hearts had trouble pumping so much blood against gravity and up such a long neck. A small brain requires less oxygen and hence less blood to carry it. Therefore, continued dinosaur evolution would not necessarily lead to a higher intelligence.

Another race of humanoid reptiles featured in the 1981 story Full Circle.

The 4th Doctor, (Tom Baker), comes across a race of humanlike aliens who have been stranded on a swamp planet. For generations they have been repairing a giant star liner take them back to their home planet Teradon. For a few months each year - in a season called mistfall - the people locked themselves into the Star Liner to avoid creatures called the Marsh Men who come out of the swamp. The Doctor finds out that the elders of the race - called the Deciders - are keeping a terrible secret from their own people. The Starline is a sham. There is no home planet Teradon. And the people have evolved from the Marsh Men themselves.

Humanity seems to have a fascination with human/reptile hybrids. The ancient Egyptians venerated Sebek - the crocodile headed God of the Nile. Biblical scholars in the past reconstructed the serpent in the Garden of Eden as a humanoid reptile with arms and legs which had been cursed by God and made to walk on its belly. Lizard-men type creatures are still reported today. One of the most dramatic happened on June 29th 1988, in Scape-Ore swamps, near Bishopville, South Carolina. Seventeen-year-old Christopher Davies was changing a flat tyre on his car, when he was attacked by a 7 ft tall, scaly humanoid figure with red eyes. It pursued his car, and leapt onto the roof, scratching at the doors and windscreen. The creature was sighted many times again that year, but then disappeared as mysteriously as it had appeared.

The season thirteen (1975) story, Terror of the Zygons, starred Tom Baker as the 4th Doctor. It pits the doctor against the Loch Ness monster. In this story the monster is depicted as a dragon-like cyborg called The Skarsen. It was brought to earth by an alien race called the Zygons, who resembled giant foetuses. The Zygons lived beneath Loch Ness in their crippled spaceship - as their home world has been destroyed. The shapeshifting Zygons intend to terraform the earth and colonise it by sending a signal to attract other members of the race. The Skarsen itself is used like an enormous watchdog and is sent out into the North Sea to destroy oil rigs (it's a dangerous job working on an oil rig in the Dr Who universe), and when the Doctor thwarts the Zygon's plans, the Skarsen returns to live in Loch Ness.

The Loch Ness monster is just one of hundreds of large, unidentified creatures, that inhabit the world's lakes. In the northern hemisphere, most of these lakes occur on, or around the 50th parallel. There have been many suggestions as to what the Loch Ness monster actually is. These include giant worms, long necked seals, and the ever-popular plesiosaur theory. In fact, none of the above are good candidates for the Loch Ness monster. Loch Ness is very cold, and only has a relatively small fish population.

Full page feature introducing Episode 1 of "Terror of the Zygons". 30 August 1975. Illustration by Frank Bellamy.

If the animal is an air-breather, it would be seen at the surface more often. The monster is likely to derive its oxygen directly from the water, and is probably some huge fish. The best candidate is some form of gigantic eel 30 ft or more in length.

Not all lake monsters can be explained as giant eels or huge fish. For example, Lake Wembu in Tibet is home to a giant reptilian creature of the size of a house. It has green scaly skin, a long neck, a large head and sharp teeth. The creature has destroyed boats and eaten fishermen. In the early 1980s it killed and ate a yak belonging to a Communist Party official who was visiting the area. So far, no scientific expeditions have visited the lake.

Another monster-haunted lake is Lake Tien-Chen, which lies on the border of northeastern China and North Korea. This huge lake was once a volcano and is said to be the home of bizarre creatures. Witnesses report something with a horse-like face, an elongated body, and black scaly skin. In 2002 the greatest mass sighting of any lake monster took place, when no less than 500 people saw the animal leaping from the water. Some of them were as close as 30 feet to the creature. It seems probable that these great creatures are some form of enormous reptile. They are quite possibly behind the legends of the Oriental dragons.

In the 1971 Doctor Who story, The Mind of Evil, renegade Time Lord, The Master, is posing as a psychologist called Dr Keller working with criminals. Using something he calls The Keller Machine, he is curing them of their criminal tendencies. The machine, in fact, contains an alien parasite that feeds on fear. It can generate images of what the victim most fears in order to create a reaction. In one sequence, the master, uses the parasite to prey on the American Ambassador's fear of China. The Ambassador's fears take on the form of a huge dragon.

In the 1979 story, The Power of Kroll, the 4th Doctor, while searching for a piece of The Key to Time, lands on a planet which is a reserve for swamp-dwelling natives which have been driven from their home world by humans. However, a company from the Earth has set up illegal gas refining stations on the swamps of the planet. The native swamp dwellers are rebelling and call upon their god Kroll to destroy the outsiders that they referred to as "Dryfoots" . Kroll reveals itself to be a gigantic octopus - a mile across. The massive size is a side-effect of the creature having swallowed a segment of the key to time.

A couple of years ago, I dealt with giant octopuses in a Yearbook article about the Cryptozoological aspects of the Biggles books. Suffice to say, that sightings off the Bahamas, and a massive carcass washed up on a beach in St Augustine, Florida, in 1897, suggest a species of octopus - known to the locals as the lusca - which has a tentacle spread of 200 ft. Not, however, having eaten a portion of the key to time, I think it is unlikely that we shall ever discover an octopus a mile across.

Just as this present volume went to press, it was announced that there will be a new se-

MYSTERIES AND MONSTERS OF THE GREAT DEEP

by Harold T. Wilkins

Mysteries and Monsters of the Great Deep

STRANGE STORIES OF SEA SERPENTS AND DEVIL FISH, AND OF THE SURPRISES THAT THE NEW SCIENCE OF THE SEA MAY HAVE IN STORE FOR US

By HAROLD T. WILKINS

HALDEMAN-JULIUS PUBLICATIONS
GIRARD, KANSAS

THIS IS A BOOK OF ENTERTAINMENT FOR HARD-BOILED SKEPTICS

Do there exist in the depths of the oceans, 3,000 fathoms below, where the ice-cold waters press with the enormous force of three tons to a square inch, or in the middle deeps of the great seas, as off the Azores, in the North Atlantic, certain marine and elusive monsters whose reality is a hoary jest in wardrooms of battleships, or staterooms and lounges and officers' messes of liners and ocean freighters, or a source of bated whispers to another set of men who voyage the wide seas in steamships and tramps, or what are left, today, of the old-time windjammers?

An old Hebrew poet who lived as far back as the days of Babylon, and whose skeptical epic has been garbled by designing priests and theologians, spoke of: "the great dragon of the waters whose teeth are terrible round about, and whose scales are his pride." He was followed by the Roman poet Claudian, who, about the 40th year of our own era, wrote of: "Britannia whose cheeks are died with woad, while a long robe of sky-blue sweeps round her ankles, and her head is covered with the skin of a Caledonian monster." (*Caledonia velata Britannia monstro*: De Laud: Stilichonis, II, 247.) This was written around 97 years after the Roman legions conquered Britain, but nobody, even today, has the ghost of an idea what "monster" was meant.

In a word, is the monstrous "sea serpent" whose existence is scouted by marine biologists and natural history museum experts anything more substantial than a pure fantasy and hallucination—matters for the psychologists or mythologists, rather than for the biologist and evolutionary scientist? Did he once exist in the Cambrian and Devonian seas, but became extinct millions of years ago, and so is the affair of the paleontologists, who may dredge up an extremely ancient bone of an extinct species and put it under a glass case in a museum?

When the lush warm years of the early ages, which must have seemed as though they would never come to an end with their hot swamps and steamy, reedy marshes and moist, green plains, slowly gave way to the oncoming glaciation, and titanic mountains rose where once there had been warm seas or shallow beaches, off which swam and hunted vast saurians and armor-plated sharks and other fearsome monsters, did *all* these monsters vanish as an unpleasant dream, leaving behind them merely a substantial wrack of fossils and petrified bones?

NO ONE HAS EVER PRODUCED A SEA SERPENT

There are two ways in which we may answer these questions, and the only conclusive way would be to produce a *whole, undamaged and living specimen*—or a dead one—of some monstrous sea serpent, or ocean saurian: the last of the great barons of the mesozoic deeps. That, so far, no one has been able to do; albeit I, myself, have photographs of monstrous-sized skulls and partly eaten or decayed creatures of no species known to modern science, and which have either been cast up on lone beaches, or dredged up from the depths of the oceans or seas, in recent years.

But what we can do is to assemble some ancient and modern testimony, which will be found, at least, entertaining, and leave the hardened skeptic to make merry over the writer who has provided him with some choice tid-bits for his scoffing jaws to chew over.

I go no farther back than 1522 of our era, 12 years before the Frenchman, Jacques Cartier, left his ship in the St. Lawrence River of eastern Canada, and talked with the Red Indian Iroquois in sight of Perce Rock. In that year, one Olaus Magnus, Archbishop of Upsala, Sweden, which is not far from the Aaland Islands, in the Baltic Sea, wrote a queer book, illustrated with some queer engravings, about a remarkable marine monster, which, he said, rose out of the waters in a Norwegian *fjord* (or long narrow arm of the sea running inland between tall cliffs). The monster stretched out his neck, reached up towards the top of the cliff, and snaffled for breakfast three sheep he found there grazing. This monster was both hungry and angry. The old archbishop swore he was 200 feet long, 20 feet round the waist, and had glaring fiery eyes and a coal-black mane. An artist of the 16th century, illustrating "the boke of Olaus"—which, by the way, was certainly *not* intended for a hoax—drew a lively picture of this monster, which he depicts in the act of rising from the sea, coiling himself round a galleon, of which he made a hearty meal, dining off sails, masts, cordage, hull, fittings, carronades, swivel guns and the skipper and crew, who had no time to protest or flee. This is the far-famed *kraken*.

Well, as pirate Long John Silver said, in quite another connection: "I would put no limits to what gentlemen (chaplings or archbishops) might consider shipshape, or might not, as the case were." So, Mr. Scoffer, don't father the responsibility of this story on the writer of this booklet; for he was neither there, nor had the honor of the archbishop's acquaintance. However, I may cite what my old friend, Roger Pocock, leader of the once-famous Legion of Frontiersmen, said a few years ago, before his death in 1940, at a hospital in Wiltshire, old England. Roger, in the 1890's, by the way, acquired fame all over U.S.A. for an epic ride, on horseback, which he made all down the Rockies far into Mexico. He told me of the "nightmare waters" surrounding the Lofoten Islands, northwest of Norway, "with their tidal overfalls and invisible islands, their awful world-famed Maelstrom . . . they are also the home of the giant cuttle-fish called the *kraken*. To the line-fishers, this creature is more than a legend." Here is a yarn told me, through an interpreter, of an adventure that befell an ancestor of Herr Nicolaissen:

"After a bad day's fishing, the skipper said: 'We'll try the Ramstakken grounds.' There, they caught cod, halibut and ling, one big fish after another, as fast as they could haul; but all the time the boat's position was changing, with regard to the mountains. The water kept on shoaling. The bottom should have been at 60 fathoms, but it shoaled until only 20 were under the keel. At 12 fathoms, the skipper said: 'Haul in! Let's row, men, as if we'd stolen boat and oars.' They had rowed about a rifle-shot away, when, on the very spot where they had been, rose arms with suckers, as long as a boat's mast, and thick as timber-stocks, which clashed together with a noise like canon-fire; and all round the sea was in a tumult . . . The *kraken* is not so often seen as the sea serpent."

Two hundred and twelve years late, (than the story of the old Archbishop Olaus), in 1734, a certain Hans Egede said he saw a monster off Greenland:

"Its head, when raised, was on a level with our maintop. Its snout was long and sharp, and it blew water much like a whale. It had large, broad paddles, or paws, and its body was covered with a scale. The skin was rough and uneven. In other respects, it was as a serpent; and, when it dived, the end of its tail, which was raised in the air, seemed a full ship's length from its body."

In 1567, Job Hortop, a mariner of Devon, England, shipped aboard the *Minion*, one of the fleet of Master John Hawkins, an old Elizabethan sea dog who combined treasure hunting, buccaneering, and galleon-clouting along with slavery, on the west coast of Africa. Hortop, after various adventures in the Spanish Main, was cast ashore with others of the ship's company in the province of Chichimeca, old Mexico. The Chichimecas sent Job up to Mexico City, where was the Spanish viceroy. Hortop, after being tortured by the monks of the Spanish Inquisition whose desire was not so much to save his protestant soul from hell fire, as to get hold of the few dollars the Lutheran sea dog had earned, laboring in Mexico, was shipped aboard the galleon of Admiral Don Juan de Velasco de Varre, bound for old Spain. As a token of their affection, the monks had clapped a yellow San Benito on his poor carcass in exchange for the dollars and pieces of eight they confiscated from him. In the latitude of Bermuda, Job and the Spanish seamen said they saw:

"A monster of the sea who shewed himself three times to us aboard the Spanish admiral's ship. We saw him three times from the middle upwards, in which parts he was proportioned like a man of the complexion of a Mullatoo, or Tawny Indian. The Generall commanded one of his clerks to put it in writing to certifye to the King of Spain."

The next yarn of a strange sea monster occurs in Spanish, in an account published in tome I of the "Museo Mexicano" of Mexico City, from which it appears that, in 1648, a "dreadful fish" came ashore on the beach of Santa Maria del Mar, Oaxaca, on the Pacific coast of Mexico. The monster scared nearly to death the quiet folk of the pueblo, who were on the point of quitting their houses to run into the interior, when a bad smell arose, and it was noticed that sea birds and dogs gathered on the shingle and began to peck and tear at the rotting carcass of the monster who had died, two days after he came ashore. He proved to have a length of 15 *varas* (41.70 feet), a height of two *varas* (5.56 feet) above the beach, and had "a red skin like that of a cow." His tail was "like a pillar" and his shoulder-blade "was shaped like a fan. . . ."

"Its ears lacked folds (*cangilones*) and it had two forefeet, but the pillared tail was so oily and greasy that not even a garbage dog would touch it. Its rib was the width of an eighth of a *vara* (about four inches) and the tail reached to the shoulder-blade and formed a striking pair of buttocks (*asiento*)."

A Spanish physician in Mexico City, Dr. Juan Nepomuceno Bolanos, examined the strange monster and made a report on him for the viceroy. A bone of the monster was preserved, as late as 1790, in the library of a Spanish convent in Mexico. Some suggest the doctor liked to hit the bottle.

About the time of the battle of Waterloo, when the combined armies of von Blucher, the German, and the Duke of Wellington, of England, out-paid to the account of the French emperor, Napoleon Bonaparte—which was in the year 1814—people in a village near Edinburgh, in old Scotland, seem to have had a hectic time when they gave thrilling chase to a strange monster that had come up out of the sea and ranged the land. The story was told in a lost manuscript unearthed, in March, 1934, by a Breton archeologist, Guy Dubois, who lives near Saint Malo, Britanny. (It is curious, by the way, that no account of this adventure appears in any contemporary Scottish, or English, journal, or magazine, which I have seen, though at that date, these old journals were by no means backward in publishing such stories):

"On April 8, 1811, an inhabitant of Edinburgh perceived, when he was on his way to Glasgow, and at some miles off the coast, an animal of extraordinary shape and size. It had a head like a bull's. The man fled as fast as his legs would carry him to the nearest village, where he roused the inhabitants. They first jeered and laughed at him, then they were incited to set forth with guns. They had not gone far before they heard an extraordinary noise, like the bellow of an animal, followed by a crash. Soon they saw the animal that made the noise. One of them went boldly ahead of his more cautious companions and got near enough to fire at it. He missed, and the monster made for him. He fled back to where his companions were in hiding. They then determined to set fire to the heather, whereupon the monster made back for the sea. The bravest of the pursuers ran after it, discharging their muskets. The first discharge caused it to stagger; the second brought it down. The weight of the monster was such that it was impossible to carry it away whole. So the villagers cut it into pieces, getting several barrels of oil from the carcass. The forelegs of the animal resembled those of a camel and its hind legs those of a horse, while it had a bull's head."

A HERO OUT-SLUGS "AN IMMENSE WORM"

In October, 1805, there was current in North Wales, a story to the effect that a small ship of the Traeth, named *Robert Ellis*, was sailing slowly on the Menai Strait, between Anglesey and the mainland in Carnarvon, North Wales. The day was calm and all was peaceful till the men on board saw what looked like "an immense worm" swimming fast after the ship. He overtook her and climbed aboard through the tiller-hole, coiling himself on deck under the mast. The crew were scared out of their lives and contemplating plunging overboard, till one man took courage and went for the monster with an oar. He and the rest drove the monster overboard, but he still followed the ship till a breeze sprang up and the ship left him in the rear.

The world was well into the age of steam and inventions, of screw and paddle-steamers and railroad locomotives, when the great monster of the deep again made his appearance. On August 6, 1848, Captain M'Quhae and three of his officers were on the decks of *H.M.S. Daedalus*, one of the early British Navy steam warships, when they saw a huge creature which they were impelled to describe in a report to the august British Board of Admiralty. I may say, here, that since M'Quhae's day, officers of the British Royal Navy sheer off reporting in their sea logs and journals about encountering what are called "sea serpents," the word having got round that the officer who does so is doing little good to himself when his name comes before their lordships of the Board, recommended for promotion. Here is the report M'Quhae made of what he saw striding the quarter-deck, when the crew were below to supper:

"We saw an enormous serpent with a head and shoulders about four feet constantly above the surface of the sea. Comparing it with the length of our main topsail yard, we estimated there was at least 60 feet of the animal on the surface of the water, and no proportion of it was, in our perception, used in propelling it through the water. ... It passed rapidly and so close under our lee quarter that had it been a man of my acquaintance, I should easily have recognized his features with the naked eye ... It did not deviate from its course to the southwest, towards which it held at the pace of from 12 to 15 miles an hour, on some determined purpose. The diameter of the serpent was about 15 or 16 inches behind the head ... Without doubt a snake's. We kept it in sight of our glasses for 20 minutes, and never once did it go below ... Its color was dark-brown, with yellowish-white at the throat ... no fins, but something like the mane of a horse, or, rather; a bunch of seaweed, washed about its back. It was seen by the quartermaster, the bo'son's mate, and the man at the wheel, besides myself and my officers."

AN EXCITING STORY THAT BELONGS IN HOLLYWOOD

But the most exciting of all stories of such encounters was that told by the skipper, Drevar, of the barque *Pauline*, who, with his crew, on July 8, 1875, said he saw one of the most amazing sights that have ever terrified tough shellbacks at sea. Says he:

"My glasses showed me a monster sea serpent coiled twice round a large sperm whale. The head and tail parts, each about 30 feet long, were acting as levers, twisting themselves and the victim around with great velocity. They sank out of sight about every two minutes, coming to the surface still revolving, and the struggles of the whale, and two other whales which were near, frantic with excitement, made the sea in this vicinity a boiling cauldron, and a loud confused noise was distinctly heard. This strange occurrence lasted some 15 minutes, and finished with the tail portion of the whale being elevated in the air, then waving backwards and forwards, and lashing the water furiously in the last death struggle, when the whole body disappeared from our view, going down head foremost towards the bottom, where, no doubt, it was gorged at the serpent's leisure.... Allowing for the two coils round the whale, I think the serpent was 160 or 170 feet long and seven to eight feet in girth. It was much in color like a conger eel, and the head, from the mouth, being always open, appeared the largest part of its body."

I shall have presently something more to say of encounters between monstrous sea saurians, or giant squids or devil fishes, and the valiant killer whale, or *orca*, who appears to feed on such monsters.

Here, again, my old friend, Roger Pocock, told me the wild story of an encounter between one, Curly Williamson, of Lerwick, about 60 miles north of Scotland, in the Shetland Ilse of Mainland:

"Curly said he had a battle with Gowdie's sea serpent, a veritable monster, in April, 1881. I have a print of the photograph of the monster's skull, dimensions being 15 feet by seven feet six inches, and weight all of three tons."

I, the writer of this booklet, have also a print of this photo, and may add that, in September, 1930, I read a press story, in a London journal, stating that the "Grimsby (Yorkshire, England) trawler, *Vedette*, landed a gigantic skull, six feet six inches long and some three feet broad, trawled from the depths of the sea. Marine experts cannot classify it, but certain characteristics suggest a whale-like appearance of a creature with two horns."

IRISH AND SCOTCH WHISKY MAY HAVE HELPED

A queer monster of this sort took a fancy to disport himself in a wild Irish loch into which tidal salt water flows. It was about 1903, and a Mr. Howard St. George was fishing at Screbe, on the wildly beautiful shores of Camus Bay, in Connemara. Said Mr. St. George:

"I saw a sea serpent calmly floating in Kilkierian Bay into which Camus Bay opens. It looked about the size of a large farm cart ... had a large hairy body with serpent's neck held erect, with head, and about six feet long (the neck)."

A few years earlier, a Scottish physician and his wife (named Farquhar Matheson) had a disturbing encounter with a monster when they were sailing in Loch Alsh, in the lone isle of Skye, in the Hebrides, of western Scotland. (I may premise the story by saying that these Scottish lochs seem to have subterranean tunnels, deep under ground, which must communicate with the sea, otherwise it is impossible to account for the appearance of unknown sea monsters who come from the deep seas off the islands. These monsters of the deep do not really inhabit Scottish or Irish lochs, or lakes. The seas around are extremely deep and little frequented, save by occasional fishermen):

"I suddenly saw rise out of the sea a long, tall neck-like thing as tall as my mast. My wife saw it, too, and was badly scared ... It was about 200 yards away *and moving towards us!* Then it began to draw its neck down, and I could clearly see it was a marine monster —of the saurian type. It had a brown and shining skin and a sort of ruffle where the head joins the neck. The head, unlike the giraffe of which, I thought, was a continuation of the neck, in a straight line."

These unknown and still uncaptured marine monsters are not all of the same species, if one may judge by the descriptions. They have been reported in widely sundered regions: from Greenland waters right down the Caribbean, and in the Pacific, off Cape Mendocino, California, also in the Indian and South Pacific Oceans, as well as the Mediterranean. Hence, it may be inferred, accepting these fantastic stories, for the sake of the argument, that they must have fathers, brothers, uncles, mothers, sisters, aunts and sons and daughters. I cite, here, a few more stories, which may prove entertaining to both skeptics and others who are inclined to a more gullible mind, though still dubious that there may be a fallacy somewhere, or theories based on incorrect observations.

A FIELD DAY FOR GULLIBLES

On May 10, 1874, there appeared in the sun, on the surface of the Sea of Bengal, India, a vast squid (?) looking like a bank of seaweed, basking on the surface of the waters. The strange object was seen off Galle, at the southern end of Ceylon, by the steamship *Strathowen*. The captain of another steamer, near by, fired a rifle at it, whereupon the strange objects, as the skipper said:

"At once made for the schooner *Pearl*, of 150 tons, jerking its way forward like an octopus, with an enormous train; a vast oblong body half the size of the vessel in length and as thick. It left a wake or train about 100 feet long. This giant squid, or kraken, seized the vessel and rolled it over on its beam ends. It got aboard, clutched the masts with great arms, and pulled down the ship by its weight. The ship rolled over and sank, only five being saved."

The skipper added:

"Monsters of this unknown species have been seen in the waters of Cochin China, floating on the sea, with arms missing, indicating that they may have an even more dreadful enemy in the deeps."

In January, 1879, the steamship *Baltimore* was steaming through the Gulf of Aden, bound for Bombay. Colonel H. W. Senior, of the Indian Army, returning from leave, was lounging on the poop deck, when he rushed to the rail and peered over. The heat was shimmering on the waters with true Aden "furnace port" intensity. He said:

"Off the starboard side, near the stern, but about three-quarters of a mile away, I saw a long, black creature darting rapidly through the water, then jumping out of it, then splashing in again, and coming rapidly nearer. I called the ship's surgeon, Dr. Hall, and he and a woman passenger all saw it. It was then about 500 yards off, with the steamship going 10 knots west. The creature sheered away when nearing the steamer's wake, and became lost in the blazing reflection of the sun on the waters. It was too rapid for me to be able to focus a telescope on it. I could not say if it had scales. Its head and neck, two feet in diameter, rose out of the water to a height of about 30 feet. The monster opened its jaws wide as it rose and closed them as it darted away for a dive. Its body was not visible at all, but must have been some way down as the surface was not disturbed much. The shape of its head was not unlike the pictures of dragons that I've seen, with a bulldog appearance on the forehead and eye-brow. When the monster had drawn its head far enough out of the water, it let it drop like a log of wood, then it darted rapidly forwards. It raised a wave-splash about 15 feet high, and had the appearance of having a sort of pair of wings on each side of its neck."

In July, 1879, a Captain Davidson of the steamship *Kiushiu Maru* said he saw another vast monster in Asiatic waters, and of a type unknown. Only a portion of its body was above water, but it was seen to seize and drag down a full-grown whale. This happened at 11:15 a.m., on July 5, 1879, and the steamship was then about nine miles off Cape Satena. The skipper and the chief officer saw, half a mile from the ship, a big whale suddenly jump out of the water. They sent below for binoculars, and when the whale again lept out of the water, they saw something holding on to its belly. The whale gave a third terrified leap, and, immediately after, a huge thing "about the girth of the mast of a Chinese junk," and snake-like in form, reared itself 30 feet in the air, remained erect for 10 seconds, and descended into the sea, upper end first, vanishing from sight with the whale, which was not again seen. Captain Davidson sent a sketch of the horrible monster to a London illustrated weekly.

In October, 1879, the British warship *H.M.S. Philomel* sighted, as the captain entered in the log:

"An amazing creature, ship lying 17 miles off Cape Zafaranna in the Gulf of Suez, at the northern end of the Red Sea, between Egyptian mainland and the peninsula of Sinai, time 5:30 p.m., on October 14, 1879. Creature lay about a mile off on our port bow. Its snout was raised 15 feet in the air; it opened its jaws and emitted columns of water. It was not a whale or any marine creature known. Spread of its jaws was 20 feet. The upper jaw was black; lower jaw gray round the mouth, but salmon-pink underneath. Back bore a dove-tail fin. It turned slowly from side to side, dipping its head every few seconds under water. It rose again in the air and suddenly vanished."

Then, in summer, 1893, the British steamship *Umfuli*, of the Natal Line, was steaming in the South Atlantic, bound from Cape Town. South Africa, to London when, about 500 yards away on the starboard bow, a large creature of snake-like appearance, suddenly rose from the sea. It moved at about 15 knots speed in the direction opposite to the ship's course and had three camel-like humps. The log states: "It looked like a 100-ton gun nearly submerged.

MORE QUEER MONSTERS

On October 30, 1896, fishermen of Lowestoft, England, came into harbor with a tale of seeing in the evening a queer monster a few miles away. Said the skipper of one trawler:

"The thing was seen by the lugger *Conquest*, of Banff, Scotland. All the crew of eight men were on deck, about 6 p.m., when a mile or so on their lee quarter they heard a loud noise like a big steamer cutting her way through the water. Looking in that direction they

they saw a huge serpent which now rose only 20 yards distant. They said the monster was fully 300 feet long (sic) and moved at about eight miles an hour. It resembled three enormous half circles in line, each being 10 feet high and 50 feet long, and there was room between each circle for the lugger to have passed. Still making the same noise, it passed close under the lugger's stern. All the men watched it for fully 15 minutes. They describe it as like a fishing-boat turned upside-down and equally large in girth."

Chinese fishermen in junks and Europeans in warships are by no means strangers to sights like these. Indeed, in 1897 and 1898, French naval officers saw monsters—not whales—blowing and spouting in these waters. The monsters, serpentine in shape, measured 75 feet long and six feet in girth. Says Professor Gruvel, of the French Museum of Natural History:

"These marine monsters of the Indo-Chinese seas advanced through the waters with undulating movements and blew out jets of combined air and spray, horizontally, and not straight up into the air. They seemed to have several lateral flippers and were colored black on the back and grey underneath. Their heads were like those of seals, but about twice as large, and their backs appeared covered with a kind of toothed comb."

We now come to our age of radio-steered 'planes, television, turbo-locomotives and atomic bombs and invisible rays, and it will be seen that even in this era seafarers have reported strange sights.

On March 11, 1911, the steamship *Ara*, homeward bound from Sierra West Africa, was about 150 miles off the coast of Senegal, West Africa, at around noon, when a botanist, named Punch, who was on deck, saw close to the ship, a strange creature, shaped like a turtle, rise above water. It had two diamond-shaped fins immediately behind the head. Punch looked and saw the creature lower its head under water and expose a section of its body. He wrote a statement, which was countersigned by other observers, including the skipper and two British colonial administrators who were on board:

"It was round . . . about 18 inches in diameter . . . of brownish-white color on the side exposed to view. Behind and submerged, we saw in the clear water, only 30 feet from the ship, another light-colored section. The captain estimated its length at 40 feet. It was moving sluggishly, and passed near us, so there was time to observe it closely."

A FAMOUS SCIENTIST LAUGHS AT SEA SERPENTS

The captain refused to enter the occurrence in his log, lest, as he said, "it give rise to trouble with my owners who don't believe in sea serpents." Ashore, the long deceased British professor, Ray Lankester, was sarcastic about the story. There are two other reports of the appearance of a monster of this type. In one case, it was seen by the yacht *Valhalla*, off the coast of Brazil, in 1905, and subsequently described in the proceedings of the London Zoological Society. In the other, a similar monster was seen off the coast of Gujerat, near Bombay, India, in 1921, when it was stated to have a head like a tortoise and an erect neck.

One British Museum has a gigantic skull, six feet six inches long, fished up by a Grimsby trawler from the deep sea, in September, 1930. To what species the monster belonged remains a mystery.

CONAN DOYLE ALSO SEES THINGS

Another monster, like the plesiosaurus, of the Cambrian age, has also been reported as seen at sea. One such monster was seen by the late Sir Arthur Conan Doyle, and he told the story in a London newspaper in February, 1928, at the time when the skeleton of a long-necked plesiosaurus, of an extinct species, was found in a quary at Harbury, Warwickshire, England. Said Conan Doyle (who also saw spirits at bunk-infested spiritualistic seances):

"I was on the bridge of an Italian liner bound from Alexandria to Athens, and my wife was with me. Off the coast of Aegina, with the famous temple of Neptune on our port bow, we saw a curious creature swimming parallel with the ship. It was about 30 feet under water. The day was calm and the water perfectly clear. The creature was about four feet long with large flippers. My wife saw it as well as I did. We both agree it was exactly like the pictures of the plesiosaurus . . . Could a turtle lose its carapace and live? A British navy captain saw a similar creature off the Irish coast."

A correspondent of mine in Queensland, Australia, tells me that what seems to have been a young or immature plesiosaurus was caught in a net off the Mudgee Beach, and says he thinks the skeleton was sent to an Australian museum. Later, in this booklet, I shall mention another report of the "existence" of a more fearsome marine monster off Hervey Bay, at Sandy Cape, half-way between Brisbane and the southern end of the Great Barrier Reef.

But, who, among the mysteries of the Great World War, of 1914-18, would ever have dreamt that a sea serpent was actually bombarded by the guns of the British warship, *H.M.S. Hilary?*

Let me say at once that the story was told in the dispassionate and coldly scientific pages of the leading British scientific journal, *Nature*, on March 22, 1930, and by Captain F. W. Dean, of the British Royal Navy (retired) who commanded *H.M.S. Hilary*, at the time. The creature remarkably resembled that seen by the steamer *Arc*, and the Earl of Crawford's yacht, *Valhalla*, mentioned earlier in this booklet. The observers were Captain Dean, his officers, and the crew aboard the warship, which was armed with six-pounders to attack German submarines found lurking off the coast of Iceland. (What their august lordships of the austere British Admiralty Board would have said, at any other time, about using shells, bought with public money, to fire on sea serpents, Captain Dean best knows.) Follows Captain Dean's statement:

"About 9 a.m., on 22 May, 1917, the warship *Hilary* was 70 miles southeast of Iceland, the day fine and clear, the Iceland mountains in sight on a flat, calm sea. We saw an object on the starboard bow. The ship was turned and headed for the object. When we were a cable's length away, about 200 yards, the creature quietly moved and we passed it on our starboard side, getting a good view. It lifted up its head once or twice, as though looking at us, and we saw the head as black and glossy, but having no ears, and in shape something like a cow. The top edge of its neck was just awash, and it curved to a semi-circle, as the creature moved its head as though to follow us with its eyes. The dorsal fin was a black equilateral triangle, rising till the apex was estimated to be four feet above water. The length of its neck was estimated by my crew to be between 20 and 28 feet, and the head had a whitish patch of flesh like that around a cow's nostrils. We fired shots at it with our six-pounders at 1,200 yards range, and must have scored a direct hit, because it submerged and was seen no more. A few days later, the ship was torpedoed and sunk, and our logs and journals went down with her."

The creature seen from the *Valhalla* also exhibited a dorsal fin rising about five feet out of the water, a long snake-like neck, terminating in a turtle-like head; but the head and part of the neck were lifted clear of the water, and were not merely floating awash.

Nature's comment was: "There seems no doubt that the observers saw a single living sea creature of unknown species." Or, as the late Sir Arthur Conan Doyle said: "The sea may yet have some surprises for us."

Two German submarine commanders, in the first World War, also reported encounters with marine monsters unknown to marine biology and zoology. Kapitan Werner-Lowisch, who became navigator of the German warship *Schleswig-Holstein*, in 1934, said he saw a marine monster, in the North Atlantic, when he was first watch officer in *Unterseeboot* in 1915. The night was clear and a moon shone. He made a note in his diary:

"Saw a sea serpent at 10 p.m., without any possibility of doubt. The creature had a longish head, scales like a crocodile, and legs with proper feet."

Herr Lowisch says the mate also saw the monster, which was "about 90 feet long," but when the captain came up from below, the monster had vanished.

NO TIME FOR FANCY CAMERA WORK

Commander Baron von Forstner, another eye witness of an encounter, was, he said in the newspaper *Deutsche Allgemeine Zeitung*, on December 19, 1933, patrolling the Atlantic, when he torpedoed the British freighter, *Iberian*, in mid-ocean, on July 30, 1915. After his submarine U.28 had sunk the steamer, he wrote:

"We heard a violent explosion 25 seconds after the *Iberian* had gone under the surface. A few seconds later a gigantic sea monster was hurled, writhing and struggling, 20 or 30 yards into the air. The monster was about 60 feet long, and shaped like a crocodile. Its head was long and pointed, and its four legs terminated in strong fins. The monster remained above water only 10 or 15 seconds, so there was no chance to take photographs."

He adds that the chief engineer and two officers of the watch also saw the monster, and that afterwards the four witnesses exchanged observations, while the engineer made a rough sketch. The baron comments that there may be rare submarine monsters of which scientists, at present, know nothing; which seems to be a reasonable supposition.

It was, however, in the year 1935 that the great sea serpent, in all his species, began to show himself all over the world, in places near frequented ports and cable and radio stations. For one thing, modern European governments, in those days before the second World War, spent a lot of money on tourist propaganda, and, no doubt, there existed among the nations a laudable desire to prove—perhaps not altogether in the interests of marine biology—that the other foreign fellow had not all the best monsters in *his* creeks and waters. But the international news agencies, giving these strange stories an airing, made a false start, in September, 1931, when a terrified fire ranger of Chapleau, Ontario, Canada, said he had seen a monster, 300 years old, which had strayed up from the ocean by way of the great lakes. Scoffers soon produced a wizened Indian, aged 100, who swore that he had hit that monster with an axe 50 years before, and had blunted the edge of the axe on the shell:

"It is a great sea turtle, 300 years old," said the Indian solemnly. Thereupon, Canadian skeptics chuckled.

Then someone produced a newspaper with a cable from England, where it was said:

"At Herne Bay, Kent, people are lining the cliffs at Beltinge, and excitedly watching a number of speed boats, traveling at 40 miles an hour, followed by motor boats, chasing for half an hour a strange monster said to be 20 feet long, which evaded capture. For some reason, none of the boats approached more than a quarter of a mile. It was first seen churning the water into foam at a spot about one-and-a-half miles beyond the pier. It traveled with an undulating motion and was said to have a smooth skin dark on top and yellow below. Bathers left the water in a panic, and hundreds of holiday-makers watched the pursuit from the esplanade."

SEA SERPENTS HELP TAKE THE DULLNESS OUT OF LIFE

Well, Herne Bay is a dull and prosaic watering-place, not many miles from where I am writing these lines, and it surely needs a little livening up.

Ranging north-to-west, between latitudes 51 degrees and 49 north, are the mysterious Monashee mountains, also called in some maps the Gold Range. They are in eastern British Columbia, and the Columbia River flows along their eastern flanks. On the west of this little known range is Lake Okanagan, where live the Inakmeep tribe of the Okanagan Indians. These Indians have centuries-old traditions that Lake Okanagan and parts of the surrounding Monashee ranges are the home of prehistoric reptiles. They know nothing of those reptiles of the Mesozoic ages, but their descriptions of the alleged reptiles singularly recall them. One of the Indians recently drew on wood a picture of one of these reptiles, which resembles a type of saurian. In August, 1933, the Okanagan monster, who the Indians believed to be dead, since he had not been seen for a long time, appeared in the waters of this lake, which are deep. He was said to have the head of a sheep and a body of prodigious size and girth. The Indians say he appears once a year and makes a noise resembling the explosions of the engine of a motor launch. Other observers reported that the monster—which, according to these stories must be a unique specimen of a fresh water saurian—has a dog-like snout and large head appendages, like the flapping ears of an African elephant. Three people said they saw him, and that he was all of 30 feet long. They swore that the monster rose to the surface, close to the lake shore, nodded his flappers, and then submerged. He rose again, and was no more seen. A white hunter who questioned the Indians about this monster was told that they called him the "Ogopo." Other Indians told old stories which they had heard round campfires and in wigwams. Said one venerable Indian:

"I am 90 years of age, and I have heard my grandfather tell of this monster. He is called the *Auck*. When I was a boy, I saw a bone, 18 inches long and 12 inches wide, from his back. Once a splendid deer was pursued into the lake and suddenly dragged down by something unseen. The water boiled like a cauldron. A white man and Indian were surrounded in their canoe by a school of these monsters at play. I saw one of them shoot a fountain of water 40 feet into the air above the lake."

This is one of the rather rare occasions when one hears that these monsters are not lone Timon and Athens hermits!

It may be noted that, usually, they appear in the dog days of summer in tropic seas, or inland water of the temperate zones; but the time came, in December, 1933, when an old dragon of the deep seemed to have said to himself: "I'll show these scoffers and jeering guys! They look on me as a seaside free exhibit to be classed with pretty Mary sunning her little belly and other charms naked, on the front at Brighton, England, and then looking dagger-eyed, as if to summon a policeman, at the men who gaze down on her nearly nude charms. The days are coming when I'll appear in public around Christmas, when their bells are ringing and not their other belles a-sunning 'emselves more'n half-naked on the beaches."

Came December, 1933, and the London newspapers began to sit up and take notice about a monster that had for some months been reported to have been seen in Loch Ness, a long narrow lake, inland, about half way along the Caledonian Canal, of Scotland. It is about 10 miles from where Moray Firth opens into the North Sea. A motor truck driver baited a barrel with skate and dogfish and tried to lure the monster. The barrel was attached to 30 yards of strong wire and the bait was hung on powerful steel hooks, but the barrel only came ashore, after drifting 10 miles up and down the lake. Motor coach parties went there; the late British Prime Minister, J. Ramsay MacDonald, spared a few days from trying to save the British Empire and unappreciative Englishmen, and motored there, after a heavenly Scottish minister took an oath in the presbytery that he had seen the monster furiously lashing his tail. A Scottish member of Parliament, wishing to cash in on the monster, urged the British Government in London to pass an act of Parliament preserving the valuable monster's life. An Englishman, who had been near the loch, aroused the indignation of Scottish hoteliers, not the most disinterested or least grasping race of guys in this *war-r-r-ld*, by scandalously asserting he (the monster) was merely a derrick that had fallen into the lake and become covered with fungi and branches of trees.

He was countered by another Englishman (?) bearing the Scottish name of Gray, who produced a print of a snapshot with his camera which: "showed a long serpentine body with a pointed muzzle and a head like that of a young cow."

The photo, which I have seen, appears genuine, but there are experts who know many camera tricks. The length of the monster was estimated at around 30 feet. Said Mr. Gray: "I took this picture on a Sunday, when I was walking along the loch shores. It was on November 26, 1933."

Naturally, in a pious country like the kingdom of auld Scotland, the taking of a picture on a Sunday aroused great indignation. For it is the land of great godliness where an old dame who, all of a-flutter, called on a reeling Scottish toper, emerging from a saloon in Edinburgh, on a Sabbath morning just after the church bells were ringing, to whistle for her lagging dog, was sternly rebuked by the bolking boozer. He fixed his red and goggling eye on her and said: "Hic . . . wad ye, wumman, hae' me break the . . . hic . . . Sawbath dee by a-whustlin' fer yer mangy tyke . . . hic?"

SEA SERPENTS ARE SEEN NOT FAR FROM SCOTCH WHISKY DISTILLERIES

A hot controversy started in faraway London about the reality of the monster. A Fellow of the Zoological Society of London arrived at the loch, on December 14, 1933, hired a motor boat and said he was going to haunt the waters for two weeks, night and day, till he saw and photographed the monster. To make sure, he would ring the loch-side with fires by day and have watchers send up night flares and fire guns to signal the monster's appearance. As he said this, cautious Cabinet (Scottish) Minister of the British Government in London, said he must know more about this monster before he asked the Royal Air Force to send up a *blimp* (balloon) and photograph him with a telephoto lens. Next day, a showman in London, who would have surely earned P. T. Barnum's good word, announced that he would give $100,000 to any who would capture and bring alive to him the real, honest-to-goodness Loch Ness Monster, who, however, had to be at least 20 feet long, and one which a consultative body of scientists, of the front rank, would certify as really prehistoric. It will be seen that that publicity hound of a showman "ra-ally wan'nt givin' away anything, sir!" He was followed by Sir Murdoch Macdonlad, M.P., haunted or "dinged to death" by fears of

Scottish hoteliers and tourist agencies lest the monster swim downstream on freshets pouring down from the snowy Scottish Highlands, and who begged the Government, or somebody, to make a grid to pen in the monster.

"WITNESSES" HAVE "SEEN" EVEN WITCHES FLYING ON BROOMSTICKS

At this interesting juncture, the writer of this booklet was cabled by a magazine in Chicago to proceed "on a rush" to Loch Ness and see what the furore was about, as somebody had been told that Scottish whisky distilleries were located near the Loch Ness. I may give a brief summary of what my diary recorded, at this time, of the monster who, by this time, had been dignified with the learned classification of "monstrosaurus Lochnessii." Sure, he was elusive! I wrote:

"This inland waterway is 22½ miles long by about one-and-a-half miles wide. Therefore, as it is 750 feet at the deepest part, a hunter who set to work to trawl this lake would have a job cut out for him. The monster is said to have appeared in this loch for the second time. This time, he has been seen, it is alleged, over a period of two and more years. More than 100 persons of both sexes, up to January 20, 1934, report having seen the monster, and their accumulative testimony reminds some folk of the impressive array of jurists, scholars, learned parsons and others in the 17th century, in both Britain and the eastern states of the U.S.A., who solemnly testified that they had seen old women witches riding on broomsticks. . . . Two witnesses, one a schoolmaster, on October, 1, 1933, the other, a motor truck driver, on December 27, 1933, assert they saw a monster with a black hump, five feet long, and three feet above water, which had a long neck and a black ridge cutting the water in a V-shaped wash, and seemed to be swimming with his head submerged. The schoolmaster adds that the monster's neck was arched like a swan's, and rose six feet out of the water and turned from side to side. The truckman adds that he rushed to the loch side and shouted, and the monster started and plunged with a swishing like a small turbine below water. Somebody remembers that the father of the present Duke of Marlborough told the Scottish inspector of fisheries, in 1872, that he and three parsons had seen in another loch (Hourn), in this same county of Inverness, a creature 96 feet long, with a flat and eyeless head and black body, which forged through the water, and, as it went, raised its back in ridges, which curved and then flattened. . . ."

A big game hunter, named Wetherell, said he had examined the remains of a sheep found on the shores of Loch Ness, while another hunter, named Paull, took plaster casts of a spoor near the sheep, which both said was that of a four-toed animal, and nine inches wide. Two more local people deposed that they had seen a strange creature, 10 feet long, in the bracken. A Scottish veterinary surgeon said he had been motorcycling at 2 a.m. on the north side of Loch Ness, on a moonlit night, on January 5, 1934, and nearly ran down "a large black monster. It turned, showed a small head on a long neck, looked at me, took fright, made two great bounds across the road, and plunged into the loch, at such speed that the water foamed and sprayed as it made violent haste to get away." According to him:

"Its body was about 20 feet long. I saw two front flippers, webbed, and behind were two other flippers. The eyes were oval-shaped just on top of the head. The tail was six feet long and rounded at the end. I jumped from my machine . . ."

Another man, who got on the radio of the British Broadcasting Corporation, said he and his wife, in a car on the lake side, met a monster with a long neck and enormous body. It moved across the road with great bounds and seemed to be carrying something like a small deer on its back. It was about 5 feet high, curved and gray like an elephant. "I accelerated quickly, for it could easily have upset the car had it attacked us."

Then came news that "ould Oireland," not to be outdone by the canny Macs, had a weird monster lashing the water with its tail, and apparently in some difficulty, in the mouth of the River Shannon, near Tarbert Island. The London *Daily Mail*, which had previously spoken of "fishing stories," now vehemently denounced all skeptics, and a well-known publisher, Sir John Murray, said he had in his private museum the skeleton of a sea monster which had been washed ashore in the Orkneys, north of Scotland, and given to his father by Lady Byron.

A PROBLEM FOR PSYCHOLOGISTS

The skeptics made a few comments, foremost being an eminent scientist and Rationalist, a nice fellow whom the present writer has personally contacted, in scientific journalism, at the Royal College of Surgeons Museum in London. He is Sir Arthur Keith. He justly pointed out that such a big beast would need to work long hours to feed himself and one would expect to find traces of his repasts. "If he came ashore, he should leave unmistakable footprints, seeing he weighs a ton. I think the footsteps found on the lake side are rather questionable, and the number and discrepancy of the witnesses and their testimony make one suppose that the Loch Ness monster is rather a problem for psychologists. He is not a thing of flesh and blood."

One point struck me, in my investigations. Assuming the existence of an unknown monster in the loch, it seems incredible that with such a length of body and enormous girth and weight, he could have swum over the weirs, even in flood, that lie between the loch and the outlet to the canal and the sea, as some investigators suggested. Are there, as there are in some parts of western Scotland, subterranean caves that communicate with the open sea from Loch Ness and Moray Firth?

CONFLICTING STORIES BY "WITNESSES"

The conflicting stories about the monster's length and girth made some folk suggest that there were two monsters in Loch Ness. They pointed out that the monster had been simultaneously reported at locations 10 miles apart. To which humorists replied that perhaps the monster was 10 miles long! Sportsmen and ghillies (attendants) deplored the fact that angling had been for the time being killed; for no one would venture out into the loch when a monster might come along at any moment and capsize the boat. Also, though Inverness police had been stationed all round the lake, they had not actually sighted the mysterious beast.

In January, 1934, Russian newspapers reported that a "serpent, 300 feet (sic) long, with a head like that of a horse" was seen off the coast, near Eupatoria, on the west coast of the Crimean (Soviet) Republic, in the Black Sea. Fishermen cut their nets adrift and fled ashore, while O.G.P.U. policemen who put out in a boat to verify the story, came back "pale and stammering." They were followed by motor boat patrols. This monster of such gargantuan length, must have been own brother to the creature, which, off Vancouver Island, in August, 1932, was said to have been seen by Major W. H. Langley, a barrister of Victoria, B. C., and his wife. It was said to have been previously sighted by Mr. F. W. Kemp, and a party, near Chatham and Discovery Islands, British Columbia. Mr. Kemp, who is an officer of the Provincial Archives of that state, made a signed statement:

"On August 10, 1932, I and my wife and son saw a mysterious something coming through the Channel between Strong Tide and Chatham Islands. Imagine my astonishment on seeing a huge creature with head out of water traveling at around four miles an hour, against the tide. It threw a considerable wash on the rocks, which gave me the idea that it was much more reptile than serpent to make such a displacement. The Channel at this point is around 500 yards wide. Swimming to the steep rocks opposite, the creature shot its head out of the water onto the rock, and moved its head from side to side, as if taking its bearings. Then fold after fold of its body appeared. At the tail-end it seemed serrated with something moving like a flail. The movements were like those of a crocodile."

One may compare this detail with the report of the German submarine commander above, in this booklet. Mr. Kemp added:

"Around its head was a sort of mane, which drifted round the body like kelp. The thing's presence changed the whole landscape. It did not belong to the present scheme of things but to the day when the world was young ... My wife and 16-year-old son ran to a point of land to get a clearer view. The monster seemed disturbed. The sea was calm and it slipped back into deep water, and in a great commotion vanished in a flash. Its speed must be terrific and its senses well developed ... Length around 80 feet, and must be terribly hard to photograph. Its head was much larger than a double sheet of newspaper. Body five feet long and shone in the sun with a bluish green color ... A year back, Major Langley saw the same, or a similar monster, also near Chatham Island. I enclose sketches."

As this peculiar monster was first seen in Cadboro Bay, B. C., in 1931, by a young fellow who was duck-shooting, it was named "Cadborosaurus." The young chap, Cyril Andrews, of Pender Harbor, said the monster suddenly rose close to his boat and snaffled a duck he had shot. Cyril rowed frantically to shore. He gave some more physiological details:

"The thing has a horse's head, but no ears or nostrils. Its eyes are in front of its head, which is about three feet wide and two long. It breathes in short pants, like a dog after a run. It is 40 (*sic*) feet long and has neither fins nor spikes (*sic*). It breaks the water with its head, as it swims."

THERE'S EVEN AN AFFIDAVIT

According to these stories, the above monster must be a young brother of the one seen by Mr. Kemp and Major Langley. In October, 1933, a lawyer of Victoria, B. C., saw him and took an affidavit that he was 80 feet long and 20 feet thick. He, his wife and friend were fishing in a quiet creek when the monster rubbed his head or hide on a rock and rushed for Chatham Island, perhaps before the lawyer had time to present him with a bill for the cost of the affidavit!

Now, as there are miles of sea studded with islands round Cadborough, B. C., there must be lots of hides-out for these monsters; so fisherman Andrews and 12 other persons also swore an affidavit that they had seen another monster round that bay. On January 9, 1934, an 11-year-old lad, Murray Jackson stood on a log boom while he and other lads were fishing in the Fraser River, B. C. They were attracted by antics of some strange creature out in mid-stream. He had his neck three feet out of water, when he suddenly dived and then the kiddies felt the boom heave under them. They reported seeing a monster "with a cow's head, two horns, ears and a green neck like stove pipe." Then three residents of Victoria, B. C., came forward and said they saw a strange monster make for a flock of gulls, which rose in a commotion from rocks and pecked him furiously. He reached his neck out of the water and grabbed one. He dived, came up, and was chased out to sea by the furious gulls. Well, no one can say that gull's flesh is tasty!

Away in Rome, the Pope was solemnly questioning pious Scottish Catholics about the Loch Ness monster, when a report came from Port of Spain, Trinidad, West Indies, about a monster called the "Huillia," which had been seen on the shores of an islet up the Ortoire River. He was said to have gone ashore and snaffled a sheep. Reports came that he was 50 feet long, as thick as a seaman's body (why a seaman?) and around 130 years old. In a rare Trinidad history book written by C. L. Joseph, in Port of Spain, in 1840, it was said that the "Huillia," in 1801, swam swiftly through water, went ashore and coiled himself round sheep and cattle. He was then said to be around 40 years old. Nobody reported whether the Pope wrote a pastoral or said a Paternoster for his suffering soul, but it was said in Port of Spain that an expedition with a snake-charmer was on the way to lull the monster to sleep. (Snake-charmer, with flute, quotha? Why not a million-dollar mammoth electric organ?)

"MONSTER AHOY!"

In March, 1934, the liner *Mauretania* was steaming through the Caribbean when startled stewards set the bells ringing, and passengers stampeded to the upper decks. High up on the mast-head a look-out had bawled: "Monster ahoy!" Firemen braved the objurgations of the chief engineer to get a peek at him through a porthole. They said he was 60 feet long and had four things like fins. The skipper said he was jet-black. The first officer said he was 65 feet long and six feet in girth. The third and fourth mates trained their binoculars on him and said he was 70 feet long, six feet wide, and had a head of two feet. Headed S.W., time 1:20 p.m. The monster was logged. The skipper said another monster had been sighted by the liner in a previous voyage of those waters. Someone did a sketch of him, which I have seen. It bestows Manchu-moustaches on him.

Again, in late July, 1934, the French liner *Cuba*, steaming 800 miles off the Azores, bound from the Gulf of Mexico, saw and logged:

"Strange beast with two dorsal humps, small head, long neck, and 100 feet long. The captain saw it jump 15 feet from the water, chasing fish."

I find that in that year, 1934, I compiled 32 reports of queer marine monsters reported seen all over the world, and in nearly every sea. I have room only for the most striking and hectic stories.

In March, fishermen, in Yorkshire ports, who had been accusing each other of hijacking their nets and catches, suddenly saw a monster "with mane and ears, and black and shining, dart under water, dive and vanish into a creek. A boat capsized as the monster dived under it." In the same month a most unearthly monster was washed ashore at Cherbourg, France. Old fishermen said he was a sea serpent; but Professor Herpin, of the Natural History Museum of Cherbourg, put him down as a "hyperodon," or beaked whale. (Yet, a photo I myself saw, showed the head of a giant tortoise rather than any species of whale.) A week later, two more mysterious marine beasts washed ashore on the same beach. They looked as if they had perished in some submarine catastrophe. Old whalers say that giant denizens of the deep, pursued almost to extinction, are taking refuge in the ice-bound waters off Greenland. One theory is that these monsters may have been upheaved in some submarine vulcanism in Davis Strait and Baffin Bay, in November, 1933. Professor Petit of the French Museum of Natural History went down to Cherbourg to see these monsters' remains. He made an autopsy, and but for the queer shape of the head—like that of a camel—and other oddities, which baffled him, he might have classified them as a selachian fish, or monstrous shark. In April, 1934, says a cable:

"A strange monster, said to be of unknown species, has been found half-buried in the sand on the Mediterranean shores of Roumani, near Port Said. It weighs 15 tons, is 48 feet long, and has a girth of 10 feet, with jaws about nine feet wide. Fifty fishermen could not budge the monster, whose skeleton is to be housed in the museum of the Cairo Zoo."

A lighthouse-keeper in Berwick Bay, borders of Scotland and England, got a thrill at dawn, when, through his binoculars, he saw a great creature, some 40 feet long, rise from the sea, with a great swirling of waters. Humps on its back stuck out of the sea. It moved with great speed and suddenly vanished before some fishing boats could get near it. Two golfers saw the monster the day before, in waters off Holy Island, and mistook him for a motor boat. Here is an item for June, 1934:

"A 30-foot monster, with girth of 30 feet, was washed ashore on Loch Kyne, Argyllshire, Scotland. Fishermen swear that it is neither shark nor whale. Its head is in front of its mouth and two huge fins are well forward. The massive tail is powerful-looking. The carcass lies in three feet of water at low tide. It is wondered if it is the monster rammed by the turbine steamer, *King George V*, three weeks ago, or is the one shot with an explosive bullet, at close range, second week of June, off Inveraray."

BASKING SHARKS TOUCH OFF A WILD YARN

The Swedish steamship, *Nordia* (1,316 tons), was off Cape Wrath, Scotland, when, from the wheelhouse, the chief officer shouted a warning about wreckage ahead. The steamship slowed down, and suddenly saw two huge fins cut the surface and there lay two monsters, side by side, basking just below water. The steamship crept slowly towards them, when they gracefully dived head first. (In this case, there is reason to suppose these were *muldonachs,* or basking sharks, which are known to reach 40 feet in length.)

Next comes an extract from the log of the Latvian motor schooner, *Elsa Croy,* of Riga:

"July 19, 1934, time 12:15 p.m. Weather calm and sunny. Off the island of Tiree, in the Gulf of the Hebrides, Scotland, we saw in the sea an unknown monster like a large lizard. He was a giant in size. Length about 50 feet and had an immensely long neck and vast mouth. Also, a long tail with fins underneath and on top. He had the body of a dray-horse. We put the ship on his course, and when he saw we were following him he turned and charged us at great speed. Our gunner fired his harpon at him, but the monster escaped with a part of the ship's rail."

In August, 1934, Dr. William Beebe, of the Tropical Research Department of the New York Zoological Society, saw from his bathysphere (or armored deep sea diving chamber) a mile deep, in the sea off Bermuda, a huge fish 20 feet long. He thought it was the largest thing seen at that depth, but hard to make out because it looked like a mass of fish looming out of the night of the deep sea. In the same month, a Canadian newspaper said:

"Two monosaurs, or sea serpents, who lived 60 millions of years ago, about the end of the Cretaceous age, have been dug from clay beds in Manitoba. They are 40 feet long and swam in the shallow seas east of the great prehistoric ocean that extended from the Gulf of Mexico to the Arctic, and cut North America in two. The dinosaurs roamed the shores and swamps bordering this vast ocean. Many remains of monosaurs have been found in the chalk beds of Kansas."

Two fishermen off the Coffe lighthouse, near Sydney, New South Wales, Australia, saw what they thought was a log four miles out at sea. Nearer up, they saw two legs a foot in diameter and 20 feet apart. Then the object turned over and vanished in a cloud of spray. It was 45 feet long.

I fancy this monster was really the giant catamary, or cuttle, called by the French, *le calmar*, a giant cephalopod. In some tropical waters it is said to reach 100 feet long! Most of its length consists of tentacles, of which it has 10, including two long ones, which look like legs. A row of suckers is attached to each tentacle, which, when raised, look like a horse's mane. Like the devil fish, and the octopus, it syphons itself rapidly backward. The *calmar* would be a vicious monster for a swimmer or diver to encounter—as he *is* encountered, with often fatal results, in the pearl beds of the Arabian Gulf, or the Timor Sea.

Later, in September, 1934, I heard that a New York underseas cameraman announced that he was on his way to the kingdom of auld Scotlan' to go 1,000 feet deep and film the Loch Ness monster. He said he was not quite sure how to locate his objective. Tut-tut! This gentleman should have no difficulty; for, on one occasion, he staged a find of an old pot of buccaneering doubloons which he had with great thoughtfulness, and desire not to cause disappointment, previously placed on the floor of Nassau Harbor, in the Bahamas. Thereby, he boosted a submarine film. However, the gentleman never got nearer Loch Ness than Manhattan.

On November 9, 1934, a fearsome monster called the *Moha-Moha*, by the black aborigines of Australia, and by Australian zoologists *Chelosaurii Lovellii*—after the Queensland schoolmarm, named Lovell, who saw it on the beach and sketched it—showed itself to shuddering Negroes, near the Great Barrier Reef. Scientists went out from Sydney to have a look at it, but whether they found it, I have not heard. Many years ago, the monster was described by an English scientist, Savile-Kent, who wrote a work on the Great Barrier Reef and the coral insects who made this tremendous wall of nature. The *Moha-Moha* is said to have been known long ago to American and French observers as a real sea serpent. Savile-Kent called it a "Chelonian sea serpent"—the *Chelonidae* being large sea turtles. The black fellows say it has legs like those of an alligator, and it seems to have first been seen in 1892, or, some say, in 1890.

Miss Lovell, in the 1930's, was schoolmarm to the children of the folk of Sandy Lighthouse, gives an "eye-witness's" description of it:

"It has a forked tail and bony rays, and its skin on head and neck is as glossy as satin. All the time it was out of the water, its great mouth gaped widely. After a quarter of an hour, I saw it put its head back into the sea, slowly closing its jaws. Its long neck moves under a carapace. I saw its head pointing out to sea, and it then rose up, putting a long, wedge-shaped fish-tail out of the water over the high shore, parallel to where I was standing, and not more than five feet from me. Giving a curious half-twist to the flukes, it sent a shower of fish into the air. The tail did not touch the sands, but was elevated in the air. I could have stood under its tail-flukes."

Miss Lovell made no sketches of this queer beast, and seven people, including two men from the Sandy Bay Lighthouse, signed a statement, attesting that:

"We saw the *Moha-Moha*, as described by Miss Lovell, when it was making for the shore of Sandy Cape, Australia, on June 9, 1890."

It looks as if we may say of these Chelonian monsters what Long John Silver, the old pirate, said of his parrot: "They lives for ever, mostly."

Some of these marine monsters of unknown species appear to have mammalian characteristics. One of them, 35 feet long, was, in 1934, brought dissected, to Prince Rupert from Henry Island, British Columbia. Dr. Neal Carter, of the Prince Rupert Dominion Fisheries Experimental Station, said it had red flesh, and was a warm-blooded mammal. The monster's head was horse-like, and it had rough skin covered with hair on the top part. The lower part had spiny quills.

GALWAY GETS THE JITTERS

It was in May, 1935, that old Galway, in Ireland, has, as the English say, a "spot of excitement." Fishermen landing, on its ancient quays, rushed into the old town, bawling and roaring that a vast monster had got entangled in their nets, out in the bay, and dragged boats, men and trawls for a long way, till—*phut!*—the nets bust to fragments. The whole town was roused by the news and people swarmed out a darned sight faster from their houses than any bell by the b-*hoy* of Father Peter of the White Friary convent had ever been known to fetch 'em to chapel and mass. Soon, the whole place was in an uproar. Seamen, harbormen, longshoremen, Irish beachcombers, hoboes and mendicants who ask for alms with a bolke and a *Benedicite,* customs officials, strolling monks, and Father Mick from parish church, men, women and children, all rushed pell-mell to the beach. Here was something that "Holy Saint Patrick" must have overlooked and "clane forgotten" when he made Ould Oireland safe for everything but snakes, monsters and Englishmen. They all stood on the beach, gaping and shading their eyes from the bright sun.

"Holy Vargin, Shamus, and what was *that?*"

A shot rang out across the bay. There came another, and another. Twelve shots in all. Tim Maloney in the lighthouse had been looking down on the sea wondering what in hell was the cause of the uproar on the shore, and why the whole town had rushed down to the beach and gazed over the bay. Then the sea heaved, and there came the very devil of a wash. From the top of his lantern tower, he sighted the back of an appalling black monster which certainly was not a submarine, sent adrift by the British Navy. Darting back into the tower, he got his rifle off the rack loaded it, and fired at the back. . . . A saurian, 48 feet long and 26 feet round the waist, jumped out of the water, thrashed the sea into a foaming whirlpool and fell back—after a time—dead! It was "seen" to have a head of vast dimensions, a "tremenjus" scaly body, and two knife-edged tails.

Fishermen kept on the wharves all day. Not a man would venture out. Fish-wives counted the beads on their blessed Hail Mary rosaries. They suspected that that monster out in the bay might have brothers, sisters, fathers or cousins, somewhere handy in Galway Bay, and out for blood and vengeance. No more fishing for them, bedad! Not a man cared to chance the fate of Jonah; for not one had any reason to anticipate a repetition of the miracle that was vouchsafed to *that* Man of God, who was not an Irishman or at least a Galwayman, anyway.

A tug got steam and went out to lay a hawser on the monster's tail-flukes. The vast carcass was towed to shore, where, on the beach, it turned the scales, they said, at four tons!

And that was not all! That very night, a fellow turned up in Galway from a pretty port further up the rugged coast. He brought the news that, two days before, some lively and lovely maidens sun-bathing in a quiet cove, where, as the American lady who wished to remove her stockings in the bank's parlor, said to the British bank manager: "No eye but God's could see me," had a rare fright and scattered towards the cliffs with screams. Behind 'em, on the rocks and sands, they left even their diminutive silk panties. A strange monster, with a calf's head and scaly body, had suddenly come up to blow and look at the girls all a-standing naked in the open air, like the statues in Lord Kilmainham's park. In two seconds, he vanished and rose again much nearer, and he now looked much as if he were choosing the tastiest bit of meat. Only he was not spry enough, that monster, or "kinda miscalculated the distance."

A BOSTON SHELLBACK SEES THINGS

Then came the turn of old *Bo-ohston*, where the Cabots don't, they say, speak even to God! Twenty-four Massachusetts shellbacks were out trawling off the coast, when something so darned odd happened that Skipper Adalbert Langhorne found it advisable to have it corroborated by no fewer than 22 men of his crew of his schooner, the *Imperator*. He said to a "nooshound":

"Ay, ye may laff, but if ye'd seen what we did, ye'd be a-layin' skeered to death with fright, as the Bible says, in yer lil' bed. We were out trawlin' some miles out to sea off Bo-ohston when, great snakes, a sea sarpin, 70 feet long, rose his durned neck and swam alongside. Yes, sir, he did! His head rose 20 feet above the water, as the mate here will testify. My men got out their swordfish harpoons, but before they could shoot off them irons, that thar hell-fire sarpin swam right out of range ... He had no fins, and his motion over the seas was like that of a water-snake."

A circumstantial story was, in June, 1935, given by Captain R. S. Culp, U.S.N. (hydrographic office), at the Merchants Exchange, San Francisco. It told about a remarkable encounter two U. S. steamships had had at sea, 20 miles off Cape Mendocino, Calif. He testified:

"June 5, 1935, 156 degrees true from Cape Mendocino, distant 20 miles, at G.M.T., 23. 27. steamers *American Hawaiian*, *Arizonian*, and Associated Oil *S.S. Paul Shoup*, Capt. D. J. La Doux, second officer, steamer, *La Brea*, says:

"A phenomenon was sighted from the fly bridges, two points off on our starboard bow. Thinking it was a derelict, I focused a telescope of six power on it. When I first sighted it, it was four feet above the water. When abeam it came up to a height of from eight to 10 feet. It was over 200 yards away when abeam. The creature's head was about the size of an engine-room ventilator, and the neck was a little smaller. The head was oblong and on the neck seemed about the size of a horse's head, but seemed about square at the lower end. The most prominent feature of the head was what seemed to be curved ridges at a position on the head that would correspond to where a horse's eyes are ... The creature now turned in the direction of the *Paul Shoup* for about half a minute and then looked in our direction. When it was about two points abaft our beam, it suddenly threw its head straight up in the air and sunk straight down. The remarkable thing about the creature was that small waves were breaking against its neck, all the time it was visible, and its rise and fall to the swell was slow, or about the same as a large steamship under way. Its color was a dirty greenish. It was visible for about four minutes altogether. It was not a sea elephant, as I have seen these before. ... Some one on the *Arizonian* may have seen the creature also. She passed directly over the spot, two miles away.

(Signed): D. J. La Doux, second officer *S.S. La Brea*."

The "monstrosaurus" now showed himself off the coast of Sussex, England, (August, 1936). Two days later, he (or it may have been a brother) almost made his bow to a lord mayor of an English city, a British member of Parliament, and an ex-cabinet minister of the British government, all of whom happened to be strolling along the cliffs of

Norfolk, near the hamlet of Eccles. The lord mayor said he was 40 feet long, and about a mile out, "skimming the waves at a great speed, I should say, 100 miles an hour. I could not see his waistline, but I'd say he was well streamlined." Then a seaman on board a ship bound for Hull, which is a bit farther north than Norfolk, leaned over the bulwarks and drew back with a fervent oath. As he later said to a pal:

"Dammee, mate, but I saw the body of a bleedin' somethin' 30 feet by four, slimy black in color with a big eel's head. He were a-layin' out on the water. No, don't tell me 'twas a bloody whale or shark, for I knows them bastards. I been at sea for 40 years and if I see many more of them, blast me but I quits the sea and goes on a farm. 'Twere a sea sarpint, and that you may lay to, mate."

The sea monster movie now shifted to the waters off Singapore in the Far East, where, on a torrid day, in August, 1936, an unknown monster, 46 feet long, was found floating, not quite dead. Malay fishermen, sons of old and bloody pirates, crept up on him. They towed the monster to a Dutch (Hollander) island, and this laid the ground for a ferocious dispute between the British and a Hollander resident, known as the "Karimoen." The British arrived from Singapore; they saw the unknown monster had some tusks 12 feet long, but that his greenish-back hide had been hacked to pieces by the Malays. In haste, the British cut off a portion of his tail, loaded it into a boat and chuffed back to base at Singapore. On their arrival, a cable was handed to them stating that the Hollander "Karimoen" had vetoed the export of old bones from islands in his control. Muttering that:

"In matters of commerce, the fault of the Dutch,
Is giving too little and asking too much,"

the indignant British scientists swore to mobilize a fleet of native boats and come to collect the rest of the corpse of what the Malays said was "a monstrous marine elephant."

A SEA SERPENT THAT CARRIED ITS OWN SIREN

In February, 1937, the passengers of the steamship, *Earl of Zetland*, had what must be a unique thrill—something they had not paid for in the fare. A strange monster with three large fins, or humps, standing six feet erect out of the water, and with a body length of 30 feet, suddenly appeared from the depths and started to race the steamship when she was off the Shetlands. Presently, the monster—who, like others of his kind, mistook the steamship for another monster—veered south at high speed, then he swung north, and one man aboard swore that, as the monster did so, he emitted a blast like that of a ship's siren!

Well, I can but challenge all the skeptical gents and the smart Alecs round the hot-dog counters to name any mammoth eel, muldonach, sea lion, swordfish, sperm whale, or vast dolphin, emerging from the mysterious deep, who gives a blast like a steamship's siren.

A few months later, and more of these elusive monsters came up to blow in Scottish waters. This time, he or, rather, they appeared to Colonel H. B. Donne, and his wife, who hail from Seend, Wiltshire, England. The colonel was cruising around in his motor yacht, the *Pandora*, though he was not hoping for anything in or out of a box, except, perhaps, a bottle of Scotch whisky. Says he:

"We were nearing the red beacon off Sealpay Island, in Skye, when we saw No. 1 monster. He had four humps, slowly waving in different directions. From time to time, a fifth hump appeared, which was perhaps a snout. It measured at least 40 feet in length. What species it was, I know not . . . Half a mile on, we saw No. 2 monster. My wife, the owner of the yacht (Mrs. Kitchener) and I drew up, and the skipper, Robert Molachlan, signed a statement of what we saw."

Came August, 1937, and the Loch Ness monster was seen by a schoolmaster and his wife. Through binoculars, they saw two five-foot humps six yards apart, like two upturned rowing boats, and a long beast with a wake like a speed boat. In the same month, a diver down in the waters off a boat slip at North Haven, Fair Isle, Lerwick, in the Orkneys, was shouted at through the speaking tube by two fellows who bade him come up, pronto, and chance the "bends," or caisson disease. They said that a big monster was coming up fast. On shore many excited people gathered, and perhaps it was the roaring of their amazed opinions on what he could be that caused him to veer sharp round and sheer off. But he was seen, all that afternoon, swimming round, till he headed for the Shetlands.

Two Roman Catholic priests, out in a small boat at the southern end of Loch Ness, saw a monster traveling at speed suddenly turn and make for their boat. They said he had a 10-foot hump and no head, but the 20-foot tail shone in the sun. They followed him, but again he turned and made for their boat, and as they had no taste for chancing the fate of Jonah—it is odd, by the way, how little practical trust and faith these clerical gentlemen have in these miracles!—they waited till he was 50 yards away, then made for the shore.

On September 21, 1938, the cargo steamship *Rimsdale*, bound from the Orkneys to Wick, North Scotland, logged the appearance of a strange "fish," when three miles off Rattray Head. The donkeyman made a sketch of it. He said it was 10 feet long in the body, had an oval head on a four-foot neck which was as thick as a man's thigh and remained in the track of the steamship for half an hour till it went under water. A similar monster had been seen off Stroma Island, not far away, in the previous week.

BIG EELS

The last marine encounter with a monster, of which I have any knowledge, happened in September, 1937, when two long-distance yachtsmen, Enno and Sylvia Loo, bound to Madagascar from Tallinn, which is the new name of Reval, the Baltic port of Esthonia, got a bit of a shock:

"Sylvia and I were steering the yacht when there came a hell of a splash like an airplane bomb falling into the sea. We were then off Berriedale Head, Caithness, Scotland. Next, believe it or not, we saw a large head peer above the water. It was two feet long. The monster passed us by, and I guess his length was all of 20 feet."

Some folk incline to think that, in some cases, these monsters may be middle-aged eels, which, having returned to spawn off the Bermudas, remain in the depths and increase in size, out of all apparent reason. Such eels, they say, may have a fancy to revisit the scenes of their infancy where they were elvers. However that may be—and if only a few of the stories we have cited have a basis of fact and truth, these monsters cannot *all* be anguilliform—the size of eel larvae, found at great depths in the sea, is astounding. In the Marine Biological Laboratory, at Kobenhavn, Denmark, there is an eel larva more than six feet in length! It was fished up off the Cape of Good hope, South Africa, and came from so great a depth that it died before reaching the surface. The Danish Marine investigation expedition, of 1930, brought home this larva in the steamship *Dana*. Professor Johannes Schmidt, leader of the expedition, calculates that had this eel larva reached maturity, the eel would have attained a length of more than 100 feet!

One may speculate whether these great creatures, such as the giant squid, and monstrous deep-sea cephalopods, whose models in wax, if seldom physiological remains, are to be seen in the maritime section of such large museums as the British Natural History Museum, at South Kensington, London, frequent the middle depths out of the reach of trawls, or are denizens of abysses in the floors of the oceans. One such ocean abyss, called by North Sea fishermen, the "Devil's Hole" was, some years ago, sought for, with echo-sounders, by the British Admiralty survey ship, *H.M.S. Fitzroy*. There had been complaints from fishermen that they had lost their gear in what seemed to be a bottomless sea not marked in the charts, or mentioned in the Sailing Directions for mariners, or the "North Sea Pilot." The *Fitzroy* found a great hole in the ocean floor which the hydrophone measured as 150 fathoms deep. Later, the ship found another chasm quite 10 miles long and deep. The chasm may mark the site of an old sea estuary existing in the far ages of long-vanished land-bridges.

Monsters of the deep seas and ocean have equally monstrous enemies, as the reader may have noted, in this booklet. For example, the gigantic killer whale, known as the *Orca gladiator*, a dreaded tiger of the sea whose jaws are filled with cruel, bristling teeth, is the foe of the horrid monster known as the Devil Fish, or the Giant Squid. The *Orca* is as valiant as he is ferocious. Indeed, he is perhaps the most formidable leviathan swimming the seas between the Horn and Baffin's Bay. A fight between the *Orca* and the giant squid of the middle deeps, who is called *Architeuthis Dux*, is a sight calculated to shake the nerves of the toughest mariner who ever sailed the seas in any age. Just imagine yourself within a deep sea cylinder, fitted with powerful searchlight and motion picture camera, as the loathsome form of *Architeuthis Dux* comes in sight, gazing with its cold and baleful eyes, which are bigger than saucers. Then, through the penumbra of the rays of the searchlight glides the vast bulk of another monster, whose powerful fins and flukes have swept it down from the surface, more than a mile above.

Like a flash, the giant *Orca* hurls its vast bulk onto the Devil Fish. A strong, curved beak projects from under the glittering saucer eyes of the giant squid, its body is sheathed with large, solid, rhomboidal scales, curving spirally like the cusps of a fir cone, over part of its body. On its powerful fins gleam luminescent spheres. Eight enormous tentacles, of immense muscular strength, leap out from the upper part of its scaly body. They are thicker than a man's arms and covered with great suckers, as well as armed with sharp claws of tiger force to rend. This is the monster of whom sailors tell strange tales. Sometimes, his withered body is found stranded on lonely beaches, but *Architeuthis Dux* is too agile, too cunning, too powerful a swimmer to be caught in nets of any deep sea survey ship. To and from dash the gigantic tentacles in the night of the great depths. Now, opening its great, bristling jaws, the enormous killer-whale seizes the muscular cephalopod with the formidable teeth of the lower cetacean jaw. Lashing in fury, the monster squid envelopes the whale's head and face with the suckers of tentacles that are 21 to 30 feet long! Shoals of small fish dart away in panic. There is a tremendous whirling and lashing and flailing, and the battle of the titans of the abyss enters the second round. Now is the time for the deep sea movie man, within the steel walls of his bathysphere, to turn his handle and shoot the scenes of the appalling war going on outside the transparent walls of thick, clear plate glass!

The killer's object is to swallow the giant squid, which is his principal food. Clashing his jaws he bites off and swallows 25 feet of tentacle. The squid makes tremendous resistance. Digging his claws deep into the killer's hide, he bites furiously with his curved beak. Tearing out great shreds of red flesh and skin, he leaves rings three feet wide, caused by the powerful suction of the cupped ends of the tentacles. Vain efforts! The killer has the monster squid clamped in a vise! The squid is gripped firmly between the bristling teeth of the killer's lower jaw and the corresponding recesses of the upper jaw. Lash and bite and rend and tear as he may, the giant squid is doomed. Struggling in frenzy, the squid is borne by the still enveloped killer-whale to the surface of the waters, far above. Here, the sea is thrashed into foam and reddened with blood. The killer tries to swallow a portion of the squid gripped in his jaws. The squid makes a last desperate attempt to save itself. It fastens its giant suckers on the whale's lip, tearing great holes in it. Torrents of blood gush forth. Suddenly, the squid's head parts from its body under the enormous tension! The body is at once gulped down by the whale, and the head drops off into the reddened sea, to be swallowed by another whale—one of a school of three *Orcas* spouting in the vicinity.

Two days later, the look-out on a steamship-whaler sights the same killer-whales three miles astern. The whaler is swung round and steams fast towards the school. The harpooner standing ready, fires his Svend Foyn harpoon, the lance smashing into the flanks of a killer. The "pram" is launched and the crew make ready to attach the carcass of the dead killer, when up flash his comrades and attack the "pram"! They maneuver around so as to squeeze it between their monster flanks. This move, the crew foils by hauling the dead killer close to the boat to serve as a shield. The rounded sides of the "pram" and the intervening carcass baffle the ferocious monsters, who succeed merely in raising the whale-boat out of the water. Now, another boat is launched from the steamer and rowed to the rescue. A lucky shot from hand-harpoon kills another *Orca*, which, opening wide its cruel jaws, in its violent death "flurry," emits the tentacles of a monstrous squid, partly digested. This battle between the killers and the whales has lasted one-and-a-half hours, and engaged the whole energies of 17 men in four boats.

Lest my readers suppose this is a fantasy of mine in the H. G. Wells and Jules Verne vein, let me quote from a learned work, in French, by the late M. Louis Joubin, an oceanographer and marine biologist who accompanied the famous deep sea expeditions of the Prince of Monaco and Monte Carlo, in 1892-1897, on the schooner *L'Hirondelle* and the finely equipped steam yacht, *La Princesse-Alice II*. Some parts of the anatomies of monstrous deep sea octopi and squids may be seen in the oceanographical museums of Monaco, on the Riviera. Dr. Jules Richard, *chef du Laboratoire de la Princesse-Alice*, has these words about the giant cephalopods of the middle deeps of the oceans:

"*Chiroteuthis Grimaldi* was taken off the Azores, doubtless when the trawl was coming up from a depth of 1,445 meters (nearly a mile deep). The transparency of its tissues and other features show that we are not here dealing with a species living on the ocean bed. This cephalopod is remarkable ... for special organs whose histological structure has led M. Joubin to essay the hypothesis that these organs serve to perceive heat rays. Each of them would be, on his hypothesis, a 'thermoscopic eye' ... The killer-whales, or cachalots, are great eaters of these monsters. They seek them out in the profound deeps of the oceans to which man has not yet succeeded in penetrating. One of these cetaceans, captured on July 18, 1885, off the Azores, by native whalers, had swallowed a collection of these monstrous and rare cephalopods. One part he belched up in dying, the rest we found in his stomach when he was cut up. Thus, we obtained *Cucioteuthis unguiculata Steenst*, of which species was found another individual lying mutilated on the surface of the sea, in 1897." (Translated from the French.)

I have not the space here to give many of the thrilling stories in my files of the life and death encounters and adventures of American and other divers and folk with these giant devil fish. But I pick out two: one October morning, in 1926, an Italian artist, Mario Manichetti, was sitting on the breakwater of the port at Leghorn, painting with pad on his knees. The day was fine, the waters calm. He was dangling his legs over the quay wall. Suddenly, he felt a grip on his leg, and looking down saw with horror that a monstrous-sized devil fish, or giant octopus, had coiled a tentacle round his ankle, and was extending its grip to his calf. Throwing away his pad, the terrified artist tried to free his leg from the horrible clutch, but the monster rose half out of the water and tried to strangle him by using its other tentacles. Yelling lustily, the artist rolled over onto the top of the stone breakwater and fought for dear life. But the devil fish had a tentacle right on his throat, before a sailor, hearing the cries, rushed up an dexterously threw a knife at the beaked head of the monster, which only then let go his hold on Manichetti. In a few minutes, the artist was freed, and the devil fish was being carried in procession to the port office.

In May, 1927, Diver A. E. Hook had gone down 50 feet to the bottom of Puget Sound, Washington, U. S., to recover some fishing tackle, when he encountered a monstrous devil fish, moving slowly along with the body of a dead man in its tentacles. Hook was armed with a knife and a shot pike pole. He was soon engaged in a thrilling fight with the monster, which attacked him, though it had, at the same time, a dead man in its tentacles. The devil fish tried to coil its tail tentacles round the diver's legs, and pike pole. Again and again, Hook stabbed at the devil fish with his pole, but before it removed its clutch and lay inert in the mud, Hook had to cut its encircling tentacles to pieces. Finally, he hauled the devil fish to the top of the water, when it was found that the dead man in its grip was a sailor who had fallen off a tug some hours earlier.

No wonder, then, that the British Army authorities at Gibraltar, Spain, stringently veto bathing by soldiers of the garrison in the waters adjoining the fortress, which are known to be the haunt of giant devil fish.

In the graphic novel "Moby Dick," in which the horror and beauty of the sea is delicately balanced with Herman Melville's New England realism, we are told how, one serene blue morning in the South Seas, the whaler *Pequod* was slowly wading through the vast sea meadows of the minute yellow brit on which the right (Arctic or Greenland bowhead) whale feeds. A stillness almost preternatural spread over the waters. The ship was holding on her way northeast from the Crozetts to Java, "the slippered waves whispering together as they softly ran on" down "the long, burnished sun glade on the water. . . . In this profound hush a strange specter was seen by Daggoo from the mast-head. In the distance, a great white mass lazily rose, and rising higher and higher, and disentangling itself from the azure at last gleamed before our prow like a snow-slide new slid from the hills. Thus glistening for a moment, it slowly subsided and sank. Then once more arose and silently gleamed."

The look-out concluded that it was "Moby Dick," the dread white whale. He yelled out: "There she breaches. Right ahead! . . . The White Whale!"

Four boats put off from the whaler. As they were pulling towards their prey, it sank, and they waited, with suspended oars, for its reappearance. . . . "Lo, in the same spot, it once more slowly rose! . . . We now gazed at the most wondrous phenomenon which the secret seas have hitherto revealed to man . . . A vast pulpy mass, furlongs in length, and breadth, of a glancing cream color, lay floating on the water, innumerable long arms radiating from its center, and curling and twisting like a nest of anacondas, as if blindly to clutch at any hapless object within its reach.

"No perceptible front or face did it have; no conceivable token of either sensation or instinct; but undulated there on the billows, an unearthly, formless, chance-like apparition of life."

It slowly disappeared with "a slow, sucking sound."

"It is the great live squid," says the mate, Starbuck.

"Few whalemen ever behold it and return alive to port to tell of it!"

Melville says the old-time whalemen (rightly!) believed it furnished the sperm (?) whale with his main or only food, and that, unlike other species, he is supplied with teeth to attack and tear it. "At times, when closely pursued, the sperm whale will disgorge what are supposed to be the detached arms of the squid; some of them, thus exhibited, exceeding 20 or 30 feet in length."

Is it possible that Melville was mistaken? The sperm whale, though it has conical teeth in the lower jaw, has no functional teeth in the upper, and it is said, usually, that this species of whale feeds on the microscopic organisms, or protozoa, called *plankton*. Perhaps, it was the *Orca*, or killer whale, also called cachalot, that the old-time whaler had in mind? But so much for "Moby Dick's" great live squid. Now glance at what follows:

A steamship, voyaging from Hamburg to Hull, England, in summer, 1927, ran into a dense fog. When the mist lifted, the look-out shouted: "Derelict dead ahead!"

The captain, in the wheelhouse, on the bridge, saw directly in front of the steamship's bows what looked like a mast, 10 or 12 feet high, standing above the water. He rushed to the engine-room telegraph to order the ship to go astern. While the bell was yet ringing down in the engine-room, the "mast" moved and swam round to the port side. Here, it re-appeared with a second "mast" curved round like an elephant's trunk of astounding girth. The weird object rose a little higher, and the passengers who had rushed up on deck saw what looked like a black body. Then the monster submerged, and the fog shut down on the sea.

Was this the giant *kraken* of the Lofotens, of which old Olaus Magnus and my old friend the late Roger Pocock spoke?

In 1928, a conference of nine nations meeting at the International Geodetic and Geophysical Union at Prague, in Czecho-Slovakia, decided to send out a survey ship to sound the great oceanic deeps of the globe. It may be there that lie the secrets, not only of submarine earthquakes but of some of these strange and mysterious marine monsters. There is the Aleutian Deep, 1,500 miles long, and five miles deep, stretching from Alaska to the South Seas; the Nares Deep, dropping down five miles into the Atlantic floor; the Java Deep, six miles down, about 145 miles southwest of Tokyo Bay. A sonic sounder, recording 14,000 soundings an hour, was used. None but an unimaginative fossil would put it beyond the genius of man to reveal the secrets of even these immense deeps before our century is over, if only we can get rid of these cataclysmic wars.

There are vast tracts of ocean which soundings have never yet plumbed. In parts of the Pacific Ocean, off the Japanese coast, there are amazingly steep-walled canyons more than 100 miles wide. At the bottom of these, even now, life is known to exist, in some form. Many of these submarine bottoms are covered with a red clay in which are particles of volcanic dust. So slow has been the secular accumulation of this red clay that it has failed to cover the mammoth-sized prehistoric sharks' teeth belonging to monsters that swept the seas of the Palaeozoic ages, say, 100 millions of years ago.

Truly we know little of the vast depths of the oceans with their mysterious creatures. The enormous volume of these great ocean deeps may be visualized by the readers when they are told that the summit of the world's highest mountain, Everest, in the Himalayas, which is 27 miles high, would fall short of the surface (east of Japan) of the tremendous Tuscarora Dee, by a mile.

Oceanographical science is but in its infancy. The sea has strange surprises yet in store for us. Science will advance and reveal and conquer them in the service of man.

17 IMPORTANT BOOKLETS BY BERTRAND RUSSELL

BERTRAND RUSSELL

Bertrand Russell, the distinguished essayist, philosopher, mathematician, logician and Freethinker, recently said he enjoyed writing booklets for E. Haldeman-Julius because he's given the fullest freedom of expression, In fact, it's only in essays written for Haldeman-Julius that Mr. Russell can give circulation to the mind-liberating thoughts he feels should be made known to the average person. In the 16 booklets listed below, Mr. Russell offers the literate a feast of reason, information, logic, wit and rollicking humor.

1. Ideas That Have Harmed Mankind. Man's Unfortunate Experiences With H's Self-Made Enemies. 25c.

2. Ideas That Have Helped Mankind. A Philosopher Looks at Man's Long History, Points to the Things That Moved Him Forward, and Shows What We Must Do if Civilization Is to Grow. 25c.

3. Is Materialism Bankrupt? Mind and Matter in Modern Science. 25c.

4. The Value of Freethought. How to Become a Truth-Seeker and Break the Chains of Mental Slavery. 25c.

5. An Outline of Intellectual Rubbish. A Hilarious Catalogue of Organized and Individual Stupidity. 25c.

6. ..iow to Read and Understand History. The Past as the Key to the Future. 25c.

7. How to Become a Philosopher, a Logician, and a Mathematician. 35c.

8. The Value of Skepticism. 25c.

9. Can Man Be Rational? 25c.

10. Is Science Superstitious? 25c.

11. Stoicism and Mental Health. 25c.

12. What Is the Soul? 25c.

13. What Can A Free Man Worship 6c.

14. Why I Am Not a Christian 6c.

15. A Liberal View of Divorce. 6c.

16. Has Religion Made Constructive Contributions to Civilization? 6c.

17. The Faith of a Rationalist. No supernatural reasons needed to make mankind. Russell. (Contains a section of "Notes and Comments," by E. Haldeman-Julius.) 25c.

THE CENTRE FOR FORTEAN ZOOLOGY

So, what is the Centre for Fortean Zoology?

We are a non profit-making organisation founded in 1992 with the aim of being a clearing house for information and coordinating research into mystery animals around the world. We also study out of place animals, rare and aberrant animal behaviour, and Zooform Phenomena; – little-understood "things" that appear to be animals, but which are in fact nothing of the sort, and not even alive (at least in the way we understand the term).

Why should I join the Centre for Fortean Zoology?

Not only are we the biggest organisation of our type in the world but - or so we like to think - we are the best. We are certainly the only truly global Cryptozoological research organisation, and we carry out our investigations using a strictly scientific set of guidelines. We are expanding all the time and looking to recruit new members to help us in our research into mysterious animals and strange creatures across the globe. Why should you join us? Because, if you are genuinely interested in trying to solve the last great mysteries of Mother Nature, there is nobody better than us with whom to do it.

What do I get if I join the Centre for Fortean Zoology?

For £12 a year, you get a four-issue subscription to our journal *Animals & Men*. Each issue contains 60 pages packed with news, articles, letters, research papers, field reports, and even a gossip column! The magazine is A5 in format with a full colour cover. You also have access to one of the world's largest collections of resource material dealing with cryptozoology and allied disciplines, and people from the CFZ membership regularly take part in fieldwork and expeditions around the world.

How is the Centre for Fortean Zoology organized?

The CFZ is managed by a three-man board of trustees, with a non-profit making trust registered with HM Government Stamp Office. The board of trustees is supported by a Permanent Directorate of full and part-time staff, and advised by a Consultancy Board of specialists - many of whom who are world-renowned experts in their particular field. We have regional representatives across the UK, the USA, and many other parts of the world, and are affiliated with other organisations whose aims and protocols mirror our own.

I am new to the subject, and although I am interested I have little practical knowledge. I don't want to feel out of my depth. What should I do?

Don't worry. We were *all* beginners once. You'll find that the people at the CFZ are friendly and approachable. We have a thriving forum on the website which is the hub of an ever-growing electronic community. You will soon find your feet. Many members of the CFZ Permanent Directorate started off as ordinary members, and now work full time chasing monsters around the world.

I have an idea for a project which isn't on your website. What do I do?

Write to us, e-mail us, or telephone us. The list of future projects on the website is not exhaustive. If you have a good idea for an investigation, please tell us. We may well be able to help.

How do I go on an expedition?

We are always looking for volunteers to join us. If you see a project that interests you, do not hesitate to get in touch with us. Under certain circumstances we can help provide funding for your trip. If you look on the future projects section of the website, you can see some of the projects that we have pencilled in for the next few years.

In 2003 and 2004 we sent three-man expeditions to Sumatra looking for Orang-Pendek - a semi-legendary bipedal ape. The same three went to Mongolia in 2005. All three members started off merely subscribers to the CFZ magazine.

Next time it could be you!

Project Kerinci, Sumatra - 2003
In search of the bipedal ape Orang Pendek

How is the Centre for Fortean Zoology funded?

We have no magic sources of income. All our funds come from donations, membership fees, works that we do for TV, radio or magazines, and sales of our publications and merchandise. We are always looking for corporate sponsorship, and other sources of revenue. If you have any ideas for fund-raising please let us know. However, unlike other cryptozoological organisations in the past, we do not live in an intellectual ivory tower. We are not afraid to get our hands dirty, and furthermore we are not one of those organisations where the membership have to raise money so that a privileged few can go on expensive foreign trips. Our research teams both in the UK and abroad, consist of a mixture of experienced and inexperienced personnel. We are truly a community, and work on the premise that the benefits of CFZ membership are open to all.

What do you do with the data you gather from your investigations and expeditions?

Reports of our investigations are published on our website as soon as they are available. Preliminary reports are posted within days of the project finishing.

Each year we publish a 200 page yearbook containing research papers and expedition reports too long to be printed in the journal. We freely circulate our information to anybody who asks for it.

Is the CFZ community purely an electronic one?

No. Each year since 2000 we have held our annual convention - the *Weird Weekend* - in Exeter. It is three days of lectures, workshops, and excursions. But most importantly it is a chance for members of the CFZ to meet each other, and to talk with the members of the permanent directorate in a relaxed and informal setting and preferably with a pint of beer in one hand. Starting this year -18-20 August 2006 - the *Weird Weekend* will be bigger and better and held in the idyllic rural location of Woolsery in North Devon.

We are hoping to start up some regional groups in both the UK and the US which will have regular meetings, work together on research projects, and maybe have a mini convention of their own.

Since relocating to North Devon in 2005 we have become ever more closely involved with other community organisations, and we hope that this trend will continue. We also work closely with Police Forces across the UK as consultants for animal mutilation cases, and during 2006 we intend to forge closer links with the coastguard and other community services. We want to work closely with those who regularly travel into the Bristol Channel, so that if the recent trend of exotic animal visitors to our coastal waters continues, we can be out there as soon as possible.

Plans are also afoot to found a Visitor's Centre in rural North Devon. This will not be open to the general public, but will provide a museum, a library and an educational resource for our members (currently over 400) across the globe. We are also planning a youth organisation which will involve children and young people in our activities.

Apart from having been the only Fortean Zoological organisation in the world to have consistently published material on all aspects of the subject for over a decade, we have achieved the following concrete results:

Disproved the myth relating to the headless so-called sea-serpent carcass of Durgan beach in Cornwall 1975

Disproved the story of the 1988 puma skull of Lustleigh Cleave

Carried out the only in-depth research ever done into mythos of the Cornish Owlman

Made the first records of a tropical species of lamprey

Made the first records of a luminous cave gnat larva in Thailand.

Discovered a possible new species of British mammal - The Beech Marten.

In 1994-6 carried out the first archival fortean zoological survey of Hong Kong.

In the year 2000, CFZ theories where confirmed when an entirely new species of lizard was found resident in Britain.

EXPEDITIONS & INVESTIGATIOINS TO DATE INCLUDE

- 1998 Puerto Rico, Florida, Mexico *(Chupacabras)*
- 1999 Nevada *(Bigfoot)*
- 2000 Thailand *(Giant Snakes called Nagas)*
- 2002 Martin Mere *(Giant catfish)*
- 2002 Cleveland *(Wallaby mutilation)*
- 2003 Bolam Lake *(BHM Reports)*
- 2003 Sumatra *(Orang Pendek)*
- 2003 Texas *(Bigfoot; Giant Snapping Turtles)*
- 2004 Sumatra *(Orang Pendek; Cigau, a Sabre-toothed cat)*
- 2004 Illinois *(Black Panthers; Cicada Swarm)*
- 2004 Texas *(Mystery Blue Dog)*
- 2004 Puerto Rico *(Chupacabras; carnivorous cave snails)*
- 2005 Belize *(Affiliate expedition for hairy dwarfs)*
- 2005 Mongolia *(Allghoi Khorkhoi aka Death Worm)*
- 2006 Gambia *(Gambo - Gambian sea monster, Ninki Nanka and the Armitage skink*
- 2006 Llangorse Lake *(Giant Pike, Giant Eels)*
- 2006 Windermere *(Lake Monster)*

To apply for a <u>FREE</u> information pack about the organisation and details of how to join, plus information on current and future projects, expeditions and events.

Send a stamp addressed envelope to:

**THE CENTRE FOR FORTEAN ZOOLOGY
MYRTLE COTTAGE, WOOLSERY,
BIDEFORD, NORTH DEVON
EX39 5QR.**

or alternatively visit our website at:
www.cfz.org.uk

Other books available from
CFZ PRESS

CFZ PRESS

THE OWLMAN AND OTHERS - 30th Anniversary Edition
Jonathan Downes - ISBN 978-1-905723-02-7 — £14.99

EASTER 1976 - Two young girls playing in the churchyard of Mawnan Old Church in southern Cornwall were frightened by what they described as a "nasty bird-man". A series of sightings that has continued to the present day. These grotesque and frightening episodes have fascinated researchers for three decades now, and one man has spent years collecting all the available evidence into a book. To mark the 30th anniversary of these sightings, Jonathan Downes, has published a special edition of his book.

DRAGONS - More than a myth?
Richard Freeman - ISBN 0-9512872-9-X — £14.99

First scientific look at dragons since 1884. It looks at dragon legends worldwide, and examines modern sightings of dragon-like creatures, as well as some of the more esoteric theories surrounding dragonkind. Dragons are discussed from a folkloric, historical and cryptozoological perspective, and Richard Freeman concludes that: "When your parents told you that dragons don't exist - they lied!"

MONSTER HUNTER
Jonathan Downes - ISBN 0-9512872-7-3 — £14.99

Jonathan Downes' long-awaited autobiography, *Monster Hunter*... Written with refreshing candour, it is the extraordinary story of an extraordinary life, in which the author crosses paths with wizards, rock stars, terrorists, and a bewildering array of mythical and not so mythical monsters, and still just about manages to emerge with his sanity intact.......

MONSTER OF THE MERE
Jonathan Downes - ISBN 0-9512872-2-2 — £12.50

It all starts on Valentine's Day 2002 when a Lancashire newspaper announces that "Something" has been attacking swans at a nature reserve in Lancashire. Eyewitnesses have reported that a giant unknown creature has been dragging fully grown swans beneath the water at Martin Mere. An intrepid team from the Exeter based Centre for Fortean Zoology, led by the author, make two trips – each of a week – to the lake and its surrounding marshlands. During their investigations they uncover a thrilling and complex web of historical fact and fancy, quasi Fortean occurrences, strange animals and even human sacrifice.

**CFZ PRESS, MYRTLE COTTAGE,
WOOLFARDISWORTHY BIDEFORD,
NORTH DEVON, EX39 5QR
www.cfz.org.uk**

Other books available from
CFZ PRESS

ONLY FOOLS AND GOATSUCKERS
Jonathan Downes - ISBN 0-9512872-3-0
£12.50

In January and February 1998 Jonathan Downes and Graham Inglis of the Centre for Fortean Zoology spent three and a half weeks in Puerto Rico, Mexico and Florida, accompanied by a film crew from UK Channel 4 TV. Their aim was to make a documentary about the terrifying chupacabra - a vampiric creature that exists somewhere in the grey area between folklore and reality. This remarkable book tells the gripping, sometimes scary, and often hilariously funny story of how the boys from the CFZ did their best to subvert the medium of contemporary TV documentary making and actually do their job.

WHILE THE CAT'S AWAY
Chris Moiser - ISBN: 0-9512872-1-4
£7.99

Over the past thirty years or so there have been numerous sightings of large exotic cats, including black leopards, pumas and lynx, in the South West of England. Former Rhodesian soldier Sam McCall moved to North Devon and became a farmer and pub owner when Rhodesia became Zimbabwe in 1980. Over the years despite many of his pub regulars having seen the "Beast of Exmoor" Sam wasn't at all sure that it existed. Then a series of happenings made him change his mind. Chris Moiser—a zoologist—is well known for his research into the mystery cats of the westcountry. This is his first novel.

CFZ EXPEDITION REPORT 2006 - GAMBIA
ISBN 1905723032
£12.50

In July 2006, The J.T.Downes memorial Gambia Expedition - a six-person team - Chris Moiser, Richard Freeman, Chris Clarke, Oll Lewis, Lisa Dowley and Suzi Marsh went to the Gambia, West Africa. They went in search of a dragon-like creature, known to the natives as `Ninki Nanka`, which has terrorized the tiny African state for generations, and has reportedly killed people as recently as the 1990s. They also went to dig up part of a beach where an amateur naturalist claims to have buried the carcass of a mysterious fifteen foot sea monster named 'Gambo', and they sought to find the Armitage's Skink (Chalcides armitagei) - a tiny lizard first described in 1922 and only rediscovered in 1989. Here, for the first time, is their story.... With an forward by Dr. Karl Shuker and introduction by Jonathan Downes.

BIG CATS IN BRITAIN YEARBOOK 2006
Edited by Mark Fraser - ISBN 978-1905723-01-0
£10.00

Big cats are said to roam the British Isles and Ireland even now as you are sitting and reading this. People from all walks of life encounter these mysterious felines on a daily basis in every nook and cranny of these two countries. Most are jet-black, some are white, some are brown, in fact big cats of every description and colour are seen by some unsuspecting person while on his or her daily business. 'Big Cats in Britain' are the largest and most active group in the British Isles and Ireland This is their first book. It contains a run-down of every known big cat sighting in the UK during 2005, together with essays by various luminaries of the British big cat research community which place the phenomenon into scientific, cultural, and historical perspective.

CFZ PRESS, MYRTLE COTTAGE, WOOLFARDISWORTHY BIDEFORD, NORTH DEVON, EX39 5QR
www.cfz.org.uk

Other books available from
CFZ PRESS

THE SMALLER MYSTERY CARNIVORES OF THE WESTCOUNTRY
Jonathan Downes - ISBN 978-1-905723-05-8

£7.99

Although much has been written in recent years about the mystery big cats which have been reported stalking Westcountry moorlands, little has been written on the subject of the smaller British mystery carnivores. This unique book redresses the balance and examines the current status in the Westcountry of three species thought to be extinct: the Wildcat, the Pine Marten and the Polecat, finding that the truth is far more exciting than the currently held scientific dogma. This book also uncovers evidence suggesting that even more exotic species of small mammal may lurk hitherto unsuspected in the countryside of Devon, Cornwall, Somerset and Dorset.

THE BLACKDOWN MYSTERY
Jonathan Downes - ISBN 978-1-905723-00-3

£7.99

Intrepid members of the CFZ are up to the challenge, and manage to entangle themselves thoroughly in the bizarre trappings of this case. This is the soft underbelly of ufology, rife with unsavory characters, plenty of drugs and booze." That sums it up quite well, we think. A new edition of the classic 1999 book by legendary fortean author Jonathan Downes. In this remarkable book, Jon weaves a complex tale of conspiracy, anti-conspiracy, quasi-conspiracy and downright lies surrounding an air-crash and alleged UFO incident in Somerset during 1996. However the story is much stranger than that. This excellent and amusing book lifts the lid off much of contemporary forteana and explains far more than it initially promises.

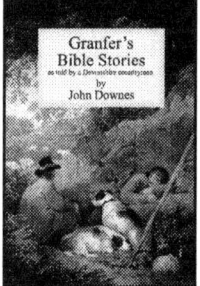

GRANFER'S BIBLE STORIES
John Downes - ISBN 0-9512872-8-1

£7.99

Bible stories in the Devonshire vernacular, each story being told by an old Devon Grandfather - 'Granfer'. These stories are now collected together in a remarkable book presenting selected parts of the Bible as one more-or-less continuous tale in short 'bite sized' stories intended for dipping into or even for bed-time reading. `Granfer` treats the biblical characters as if they were simple country folk living in the next village. Many of the stories are treated with a degree of bucolic humour and kindly irreverence, which not only gives the reader an opportunity to re-evaluate familiar tales in a new light, but do so in both an entertaining and a spiritually uplifting manner.

FRAGRANT HARBOURS DISTANT RIVERS
John Downes - ISBN 0-9512872-5-7

£12.50

Many excellent books have been written about Africa during the second half of the 19th Century, but this one is unique in that it presents the stories of a dozen different people, whose interlinked lives and achievements have as many nuances as any contemporary soap opera. It explains how the events in China and Hong Kong which surrounded the Opium Wars, intimately effected the events in Africa which take up the majority of this book. The author served in the Colonial Service in Nigeria and Hong Kong, during which he found himself following in the footsteps of one of the main characters in this book; Frederick Lugard – the architect of modern Nigeria.

CFZ PRESS, MYRTLE COTTAGE, WOOLFARDISWORTHY BIDEFORD, NORTH DEVON, EX39 5QR
w w w . c f z . o r g . u k

Other books available from
CFZ PRESS

ANIMALS & MEN - Issues 1 - 5 - In the Beginning
Edited by Jonathan Downes - ISBN 0-9512872-6-5

£12.50

At the beginning of the 21st Century monsters still roam the remote, and sometimes not so remote, corners of our planet. It is our job to search for them. The Centre for Fortean Zoology [CFZ] is the only professional, scientific and full-time organisation in the world dedicated to cryptozoology - the study of unknown animals. Since 1992 the CFZ has carried out an unparalleled programme of research and investigation all over the world. We have carried out expeditions to Sumatra (2003 and 2004), Mongolia (2005), Puerto Rico (1998 and 2004), Mexico (1998), Thailand (2000), Florida (1998), Nevada (1999 and 2003), Texas (2003 and 2004), and Illinois (2004). An introductory essay by Jonathan Downes, notes putting each issue into a historical perspective, and a history of the CFZ.

ANIMALS & MEN - Issues 6 - 10 - The Number of the Beast
Edited by Jonathan Downes - ISBN 978-1-905723-06-5

£12.50

At the beginning of the 21st Century monsters still roam the remote, and sometimes not so remote, corners of our planet. It is our job to search for them. The Centre for Fortean Zoology [CFZ] is the only professional, scientific and full-time organisation in the world dedicated to cryptozoology - the study of unknown animals. Since 1992 the CFZ has carried out an unparalleled programme of research and investigation all over the world. We have carried out expeditions to Sumatra (2003 and 2004), Mongolia (2005), Puerto Rico (1998 and 2004), Mexico (1998), Thailand (2000), Florida (1998), Nevada (1999 and 2003), Texas (2003 and 2004), and Illinois (2004). Preface by Mark North and an introductory essay by Jonathan Downes, notes putting each issue into a historical perspective, and a history of the CFZ.

BIG BIRD! Modern Sightings of Flying Monsters

Ken Gerhard - ISBN 978-1-905723-08-9

£7.99

Today, from all over the dusty U.S. / Mexican border come hair-raising stories of modern day encounters with winged monsters of immense size and terrifying appearance. Further field sightings of similar creatures are recorded from all around the globe. The Kongamato of Africa, the Ropen of New Guinea and many others. What lies behind these weird tales? Ken Gerhard is in pole position to find out. A native Texan, he lives in the homeland of the monster some call 'Big Bird'. Cryptozoologist, author, adventurer, and gothic musician Ken is a larger than life character as amazing as the Big Bird itself. Ken's scholarly work is the first of its kind. The research and fieldwork involved are indeed impressive. On the track of the monster, Ken uncovers cases of animal mutilations, attacks on humans and mounting evidence of a stunning zoological discovery ignored by mainstream science. Something incredible awaits us on the broad desert horizon. Keep watching the skies!

STRENGTH THROUGH KOI
They saved Hitler's Koi and other stories

£7.99

Jonathan Downes - ISBN 978-1-905723-04-1

Strength through Koi is a book of short stories - some of them true, some of them less so - by noted cryptozoologist and raconteur Jonathan Downes. Very funny in parts, this book is highly recommended for anyone with even a passing interest in aquaculture.

CFZ PRESS, MYRTLE COTTAGE, WOOLFARDISWORTHY BIDEFORD, NORTH DEVON, EX39 5QR

Other books available from
CFZ PRESS

BIG CATS IN BRITAIN YEARBOOK 2007
Edited by Mark Fraser - ISBN 978-1-905723-09-6

£12.50

Big cats are said to roam the British Isles and Ireland even now as you are sitting and reading this. People from all walks of life encounter these mysterious felines on a daily basis in every nook and cranny of these two countries. Most are jet-black, some are white, some are brown, in fact big cats of every description and colour are seen by some unsuspecting person while on his or her daily business. 'Big Cats in Britain' are the largest and most active group in the British Isles and Ireland This is their first book. It contains a run-down of every known big cat sighting in the UK during 2006, together with essays by various luminaries of the British big cat research community which place the phenomenon into scientific, cultural, and historical perspective.

CAT FLAPS! Northern Mystery Cats
Andy Roberts - ISBN 978-1-905723-11-9

£6.99

Of all Britain's mystery beasts, the alien big cats are the most renowned. In recent years the notoriety of these uncatchable, out-of-place predators have eclipsed even the Loch Ness Monster. They slink from the shadows to terrorise a community, and then, as often as not, vanish like ghosts. But now film, photographs, livestock kills, and paw prints show that we can no longer deny the existence of these once-legendary beasts. Here then is a case-study, a true lost classic of Fortean research by one of the country's most respected researchers; Andy Roberts. Cat Flaps! is the product of many years of research and field work in the 1970s and 80s, an odyssey through the phantom felids of the North East of England. Follow Andy on his flat cap safari as he trails such creatures as the 'Whitby lynx', the 'Harrogate panther', and the 'Durham puma'. Written with humour, intelligence, and a healthy dose of scepticism, Cat Flaps! is a book that deserves a place on the bookshelf of every cryptozoologist.

CENTRE FOR FORTEAN ZOOLOGY 2007 YEARBOOK
Edited by Jonathan Downes and Richard Freeman
ISBN 978-1-905723-14-0

£12.50

The Centre For Fortean Zoology Yearbook is a collection of papers and essays too long and detailed for publication in the CFZ Journal Animals & Men. With contributions from both well-known researchers, and relative newcomers to the field, the Yearbook provides a forum where new theories can be expounded, and work on little-known cryptids discussed.

CFZ PRESS, MYRTLE COTTAGE,
WOOLFARDISWORTHY BIDEFORD,
NORTH DEVON, EX39 5QR
w w w . c f z . o r g . u k

www.ingramcontent.com/pod-product-compliance
Lightning Source LLC
Chambersburg PA
CBHW062208080426
42734CB00010B/1837